TRANSACTIONS OF THE
AMERICAN PHILOSOPHICAL SOCIETY
HELD AT PHILADELPHIA
FOR PROMOTING USEFUL KNOWLEDGE

VOLUME 70, PART 7 · 1980

An Unusual Newly Discovered Archosaur From the Upper Triassic of Virginia, U.S.A.

ROBERT E. WEEMS
DEPARTMENT OF GEOLOGY, GEORGE WASHINGTON UNIVERSITY

THE AMERICAN PHILOSOPHICAL SOCIETY
INDEPENDENCE SQUARE: PHILADELPHIA

October, 1980

Library of Congress Catalog
Card Number 79-54275
International Standard Book Number 0-87169-707-6
US ISSN 0065-9746

PREFACE

A new archosaurian reptile, *Doswellia kaltenbachi* nov. gen. et sp. from Upper Triassic (Middle Carnian) strata was found in the Taylorsville Basin (Falling Creek Member, Doswell Formation, Newark Group) of east-central Virginia. The description presented in this study is based on a nearly complete axial skeleton, lacking most of the limbs, which is sufficiently complete to allow a thorough taxonomic review of the animal and to permit an analysis of its functional anatomy and speculations on its habits and ecological role. *Doswellia* is quite different from any previously described archosaur and as such merits placement in it own suborder (Doswelliina, new suborder) and family (Doswelliidae, new family). Its closest affinities lie with the Proterosuchia, from which it was apparently derived, but it has diverged widely from that group in the acquisition of the following traits: dorsal and lateral (but not ventral) armor, a sharply downflexed and wide tail, a heavily braced sacrum with ilia prominently distally outflared, a unique manner of rib-vertebra articulation unknown in any other reptile, and numerous features in the skull similar to specializations found in the Crocodilia but obviously evolved independently of that group. *Doswellia's* functional specializations suggest that it was a foraging animal with powerful feet and jaws suited for rooting and grubbing. The curious tail shape and structure suggest *Doswellia* could pull its tail under its body, forming an armor shield for its rear and abdomen. If this is so, *Doswellia* is the oldest known example of an armadillo-like animal.

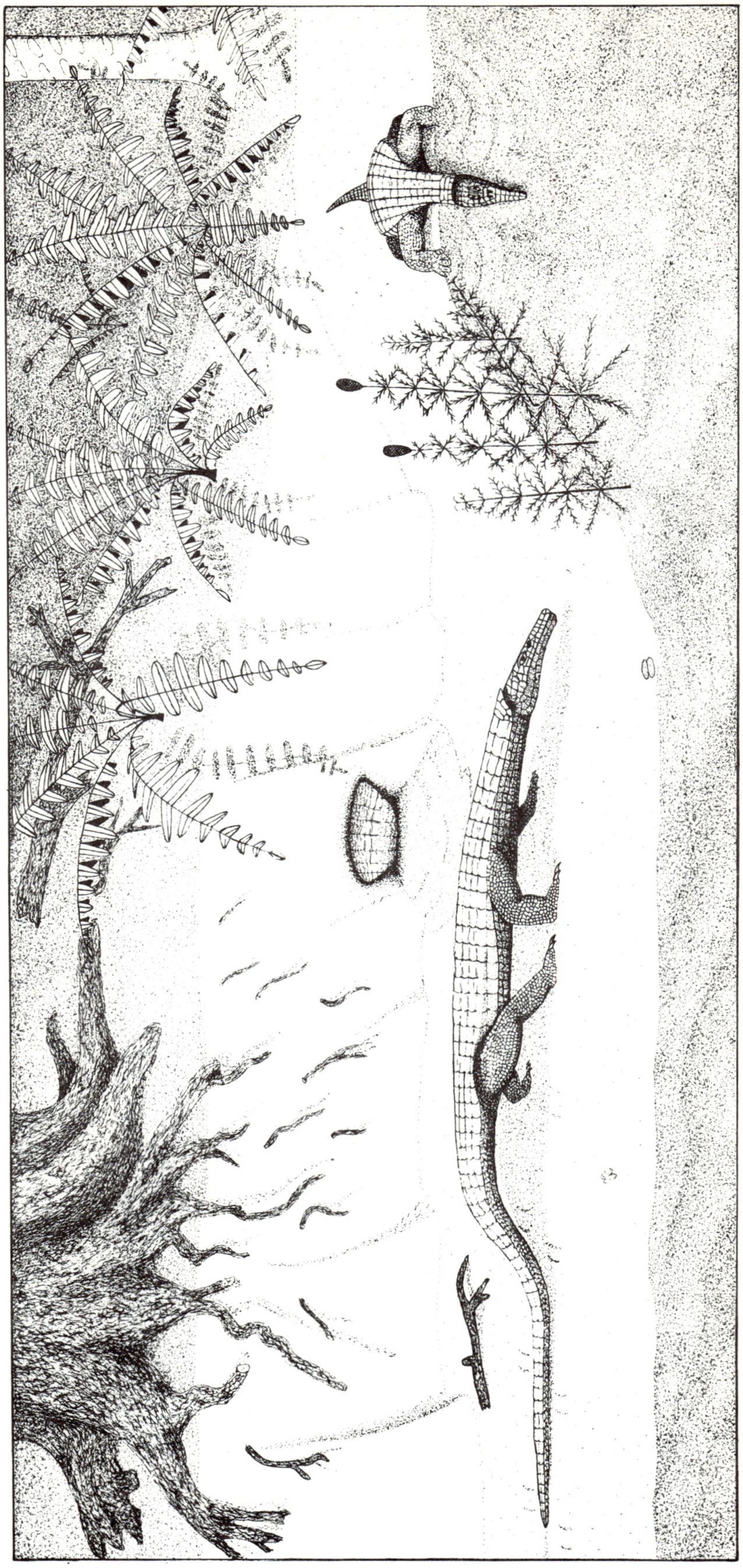

FIG. 1. Restoration of *Doswellia kaltenbachi* showing its probable appearance and probable habitat. Note one individual (center) has dug into the stream bank and pulled its tail in beneath its body.

AN UNUSUAL NEWLY DISCOVERED ARCHOSAUR FROM THE UPPER TRIASSIC OF VIRGINIA, U.S.A.

ROBERT E. WEEMS*

CONTENTS

Introduction	5
Age of the deposit and geologic setting	5
Associated flora and fauna	8
Taphonomy	9
Systematics	10
Skull	12
Mandible	17
Teeth	18
Vertebrae	18
Chevrons	22
Ribs	23
Gastralia	26
Pectoral girdle	26
Pelvic girdle	27
Limbs	27
Armor	30
Taxonomic affinities	33
Diagnosis of the suborder Doswelliina and the family Doswelliidae	37
Functional anatomy	37
Speculations on the ecology of *Doswellia*	40
Acknowledgments	41
References	41

INTRODUCTION

Most of the material described in this study was discovered and collected in 1974 during a routine examination of a sewer plant construction site southeast of Doswell, Virginia, near the confluence of the North Anna and Little Rivers (figs. 2 and 3) by a party composed of James Kaltenbach and Robert Weems of the George Washington University and Dr. Nicholas Hotton III of the Smithsonian Institution. At that time the site consisted of several large excavations which completely pierced a Pleistocene gravel deposit and exposed lacustrine to deltaic sediments of the Upper Triassic Falling Creek Member of the Doswell Formation. Kaltenbach discovered a large block which contained an articulated string of ten vertebrae and had been displaced from the wall of the largest pit. The spot from which the block had come was readily ascertained and digging soon produced more bone-bearing blocks which could be attached to the main block. This material was removed and taken to the Smithsonian for preparation.

After the bulk of the skeleton had been removed, Mr. Donald Beazely of the construction firm building the sewer plant presented the author with two slabs from the same pit, one of which contained a vertebra and associated rib and the other the back of the skull, jaws and parts of four neck vertebrae. These fragments were catalogued separately because they could not be fitted to the main blocks, but there is no doubt that the bones they contain represent the same individual. The exact lithology of all of these blocks (gray, calcareous siltstone lying above gray, calcareous clay) occurs in only one horizon (fig. 5 and table 1) so there is no question that all of these blocks came from the same bedding plane in the north wall of the pit. Furthermore, there is no doubt that all of the blocks contain remains of the same individual because there is no duplication of elements among the blocks and, despite search, no other articulated materials were found along the approximately 30 feet of this bedding plane.

During preparation at the Smithsonian laboratory the articulated material in the largest block proved to contain, as parts of a single individual, elements identical with vertebrae and armor collected just north of Ashland, Virginia, along the Richmond, Fredericksburg, and Potomac Railway property (fig. 3) in the late 1950's and 1968. This material had been tentatively assigned to *Spinosuchus caseanus* VonHuene and *Rutiodon carolinensis* Emmons in the belief that it could not possibly belong to a single type of animal. With the discovery of the skeleton from the Doswell sewer pit, however, it became clear that all of this material pertained to an entirely new and unusual reptile, previously unrecognized in the fossil record.

AGE OF THE DEPOSIT AND GEOLOGICAL SETTING

All known remains of *Doswellia* have come from the Falling Creek Member, the middle Member of the tripartite Doswell Formation (fig. 4). Palynological correlations by Cornet (1977) indicate that the age of the Falling Creek Member is middle Carnian, as is the underlying Stagg Creek Member and the basal portion of the overlying Newfound Member. No pollen assemblages are known yet from the upper part of the Newfound Member, but those low in that unit, on the basis of interbasin correlations, are suggestive of very latest middle Carnian age. Since *Doswellia* came from near the top of the Falling Creek Member, it is thus inferred that it lived during late middle Carnian time (*ca.* 210 million years ago).

The two localities which have yielded remains of *Doswellia* are in the uppermost 400 feet of the 1,200 foot thick Falling Creek Member (figs. 3 and 4). This Member consists of sandstone, shale, mudstone, conglomerate, lime-cemented sandstone and mudstone and coal which, collectively, is suggestive of paludal to lacustrine environments. The two sites are at most a few hundred stratigraphic feet apart and could be nearly equivalent in age; absence of marker beds makes a precise determination of relative position in the formation impossible. Although the Ashland site

* U.S. Geological Survey Stop 926, Reston, Virginia 22092.

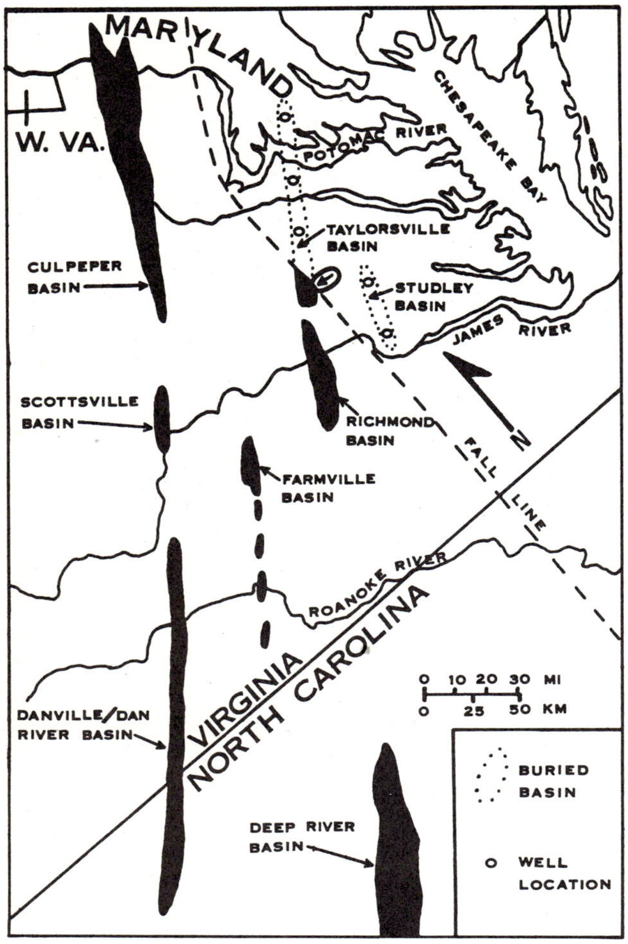

FIG. 2. Map of eastern Virginia and adjoining states showing locations of Newark Supergroup (Upper Triassic-Lower Jurassic) basins (Olsen, 1978). Remains of *Doswellia* came from the Taylorsville Basin in east-central Virginia. Subsurface basin data from open files of the Virginia Division of Mineral Resources. Arrow shows location of figure 3.

is composed only of poorly exposed green to gray siltstone with occasional bivalve casts, the Doswell site was well exposed during construction and has been described in some detail (fig. 5 and table 1). Relative proportions of carbonate, sand/pebble and silt/clay in each recognized bed are shown in fig. 6. X-ray diffraction analysis did not reveal the presence of dolomite in any bed, but did reveal the presence of gypsum and pyrite. The clay fraction similarly includes illite, kaolinite and montmorillonite; chlorite was not detected. Montmorillonite was especially abundant in the clayballs which expanded readily after being dried out. Both thin sections and X-ray diffraction indicated considerable quantities of microcline, orthoclase and plagioclase in the sand fraction, even though this was not obvious in most hand samples.

The presence of montmorillonite and the absence of chlorite, while not conclusive, does suggest a nonmarine depositional environment (Riley and Chester, 1971, p. 340). Similarly, the presence of woody material, spores and pollen, unionid bivalves, fish scales, parasuchian teeth and rauisuchid(?) bones, especially when considered along with the absence of clearly marine fossils, strongly indicate this is a nonmarine, lacustrine to fluviatile sequence. The abundance and diversity of spores and pollens indicate the local climate was generally equable and moist (Bruce Cornet, written communication, 1976). However, evaporation potential must have been high around the margins of the lake, at least seasonally, to permit the precipitation of calcite and gypsum.

The environment of bed 1 (fig. 5 and table 1) was probably lacustrine since it consists of persistently flaggy, planar bedded sandstones with occasional tree stem fragments and fish scales. Unlike beds 3–7, mica flakes in this unit are generally crushed and warped, indicating compaction prior to cementation. The lack of early calcite cementation possibly reflects accumulation in somewhat deeper water than the overlying beds. The tops of each flaggy layer in bed 1 are marked by an accumulation of large mica flakes, indicating either graded bedding or else pulses of rapid sedimentation followed by long intervals of quiescence. The calcareous clay bed (bed 2) is sparsely laminated

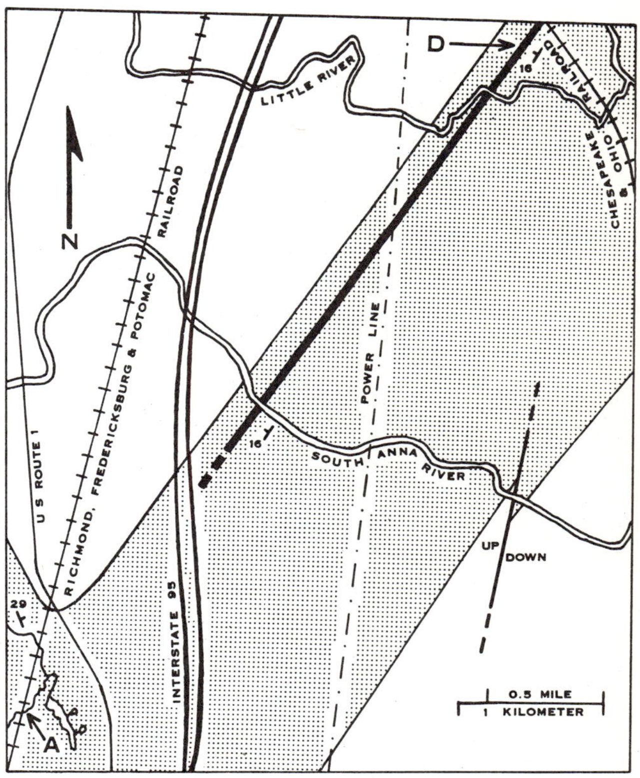

FIG. 3. Location of the Ashland site (A) and Doswell site (D) at which remains of *Doswellia* were recovered. The carbonate-rich interval in which *Doswellia* was found at the Doswell site could be traced to the South Anna River, but to the southwest outcrop is too poor to determine if it continues beyond this point. Stippled area is outcrop belt of the Falling Creek Member of the Doswell Formation.

and suggests accumulation in quieter water than bed 1. The presence of the partially articulated skeleton of *Doswellia* at the top of this bed also suggests quiet conditions of accumulation. The lower part of the overlying, unlaminated bed 3 is much less clayey and more calcareous than bed 2 but it becomes increasingly more clayey upward and less calcareous. It is, however, persistently more sandy than bed 2, containing 25–30 per cent sand. Scattered fish scales, bivalve casts, parasuchian teeth, *Doswellia* bones and rauisuchid(?) bones occur throughout this bed. Since the sand fraction is fine and the calcite is micritic, quiet conditions of accumulation are indicated. The abundance of calcite suggests shallow waters.

Bed 4 is a pebbly sandstone with sparse *Doswellia* bones, parasuchian teeth and fish scales, numerous mudballs and little calcite. This appears to be a typical channel deposit. Presence of fresh feldspar grains and granitic rock fragments indicate a close source area to the east (the only direction in which granite occurs) and the coarseness and freshness of these sediments suggest a steep stream gradient. Bed 5 is similar except that there is much more calcite and bone present and less silt and clay. The higher concentration of bone and calcite suggests deposition may have been slower than when bed 4 accumulated.

Bed 6 is much finer grained than the beds immediately below it. A few fish scales are present but otherwise no fossils were found. In bed 7 the sand is somewhat coarser and a few coprolites occur. A few fragmentary rauisuchid(?) bones, *Doswellia* bones and parasuchian teeth are present.

With bed 8 black shale deposition commenced. This shale is poorly laminated and has numerous fish scales, *Doswellia* bones and parasuchian teeth scattered throughout it. It contains only slightly less calcite than underlying beds. This bed grades upwards into bed 9 which is a thinly laminated, lime-cemented mudstone which may have been an algal mat. Bed 10 differs from 9 only in that there is a much higher carbon and silt/clay content and a somewhat lower calcite content; problematic, algal laminae remain essentially the same. Bed 11 has numerous casts of disarticulated bivalves scattered between algal mat layers. Both the shells and the calcareous interlaminations in the algal mats have been leached, yet this layer has a significant amount of gypsum (9 per cent by weight). The gypsum is presumed to be the result of some sort of secondary alteration, before or after burial, since a process which could leach out calcite but not the even more soluble gypsum is difficult to envision. The combination of algal mat laminae and scattered bivalve shells are strongly suggestive of a mudflat environment on which storm deposits of shells accumulated.

The flaggy layers of bed 1 suggest a prodelta to lacustrine depositional environment. From this bed upward a generally shallowing trend occurs through bed 5, followed by a deepening trend in 6–8. Beds 9–11 record a second shallowing trend. The quiet conditions of beds 2 and 3, with their probably bioturbated textures and high calcite content, suggest a shallow aerobic environment such as in a shallow restricted body of water in which little deposition and high evaporation took place. Possibly a prograding delta tongue cut off a portion of the lake, thereby forming a quiet lagoonal body into which animals might wander or be washed during floods. Alternatively, a broad shallow shelf along the lake margin could produce a similar environment. After deposition of bed 3, a major distributary channel migrated across the area, leaving mudball-rich deposits. Deepening of the lake or subsidence of its floor led to less

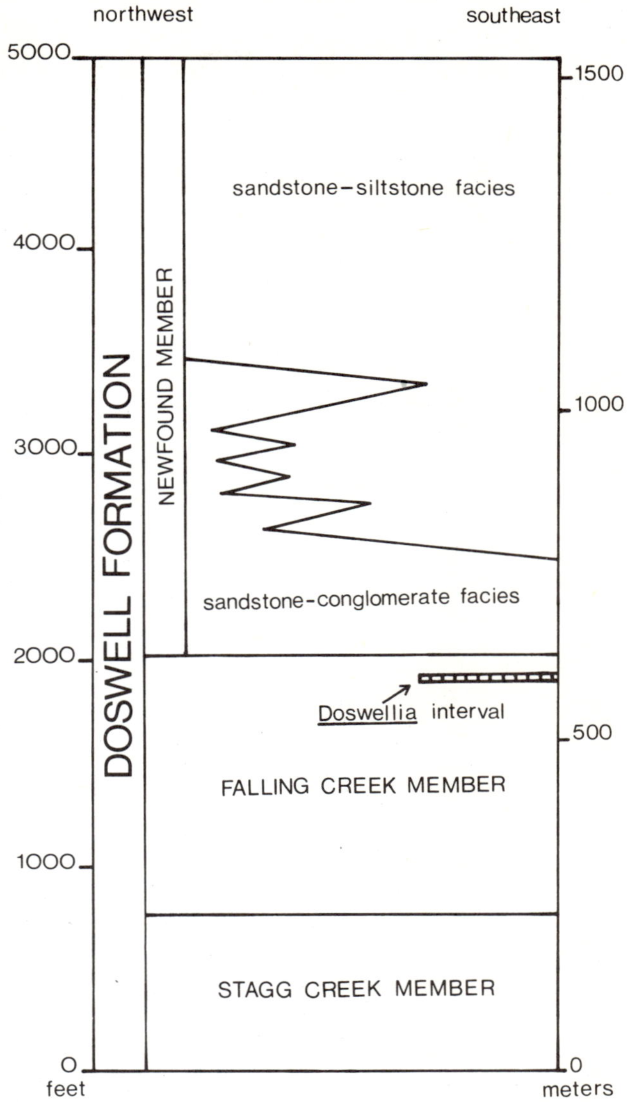

FIG. 4. Stratigraphic column of the Doswell Formation. The Falling Creek Member largely consists of well laminated sandstones, with lesser quantities of siltstones, shales, conglomerates, and carbonate-rich sandstones or siltstones.

TABLE 1

SECTION OF UPPER FALLING CREEK MEMBER EXPOSED IN CONSTRUCTION PIT AT DOSWELL SEWER PLANT.

Bed	Lithology	Thickness (Meters)
11	Shale, dark gray, gypsiferous and carbonaceous, friable; well laminated, algal mat texture; casts and molds of disarticulated clam shells numerous and thickly covering most bedding surfaces.	0.15+
10	Shale, dark gray, calcareous, silty and carbonaceous; moderately to well laminated, algal mat texture; coprolites abundant, fish scales common, clam shell casts rare.	0.30
9	Mudstone, dark gray, calcite cemented, carbonaceous; occasional veins of calcite and pyrite; moderately to well laminated, algal mat texture; occasional fish scales and bones.	0.05
8	Shale, dark gray, highly calcareous and silty; poorly laminated; numerous fish bones and scales; reptile bones and teeth.	0.45
7	Sandstone, medium gray, medium to fine grained, poorly sorted, calcareous, massive; faintly laminated; rare 2–3 mm clayballs or coprolites, numerous fish scales, occasional reptile bones and teeth.	0.20
6	Sandstone, medium gray, fine grained, moderately sorted, calcareous, massive; occasional fish scales.	0.20
5	Sandstone, medium gray, coarse to very coarse grained, poorly sorted, highly calcareous, massive; numerous 2–5 mm clayballs, occasional quartz and feldspar granules and pebbles, becoming finer upward; numerous bones, bone fragments, and fish scales.	0.10
4	Sandstone, medium gray, coarse to very coarse grained, poorly sorted, arkosic, calcareous at base but only faintly calcareous elsewhere, massive; numerous 2–8 mm clayballs, scattered quartz and feldspar granules; occasional bones, bone fragments and fish scales.	0.85
3	Siltstone, medium gray, sandy, very fine grained at base to medium grained at top, poorly sorted, highly calcareous, massive; scattered 2–6 mm mudballs and 2–3 mm muscovite flakes in upper part; occasional plant stem casts, numerous clam shell casts, fish scales, and reptile bones; *Doswellia* type and paratype specimens found at basal contact of this bed.	0.60
2	Clay, medium gray, calcareous and silty, moderately well laminated; weathers to nodular lumps.	0.40
1	Sandstone, medium gray, very fine to fine grained, moderately sorted, sparsely calcareous, faintly laminated; ½–1 mm mudballs or coprolites, occasional crushed muscovite flakes; occasional fish scales; partings developed at about 10 cm intervals on which large (2–3 mm) muscovite flakes are concentrated as well as occasional leaf and stem impressions (calamitean?).	4.00+
	Total	7.30+

energetic and more typically lacustrine conditions which produced beds 6–8. After deposition of bed 8, algal mats and calcite interlaminae accumulated in very shallow water. This finally built up to lake level to produce mudflats on which shells accumulated as storm deposits.

Although this succession of beds has not been found elsewhere, a highly calcareous succession is exposed on the Little River and South Anna River at the same stratigraphic horizon, along strike to the southwest (fig. 3). Calcite content of this magnitude is not known anywhere else in the basin, so this interval serves as a marker horizon within the Falling Creek Member between the South Anna and North Anna Rivers. Whether or not this marker interval extends southwest of the South Anna River is undetermined; exposure there is so poor that this interval could easily have been missed even if it were present.

ASSOCIATED FLORA AND FAUNA

Archosaurs, fish, branchiopids, a snail, unionid clams and plant fragments have been collected from the Falling Creek Member, but to date little of this material has been fully identified or described. Much of it is not identifiable even as low as the family level, but a list is useful because of the ecological significance of the association. At least two other archosaurs in addition to *Doswellia* occur at the Doswell locality. One, represented by two fragmentary vertebrae and the proximal end of a humerus, is possibly a rauisuchid (John Zawiskie, personal communication, 1980). The other archosaur, apparently a parasuchian, is represented by anterior piercing teeth and posterior blade-like shearing teeth, none of which are fluted. The teeth are of a size consistent with *Rutiodon carolinensis,* but in the absence of fluting a positive assignment is not possible; the material could as easily pertain to *Clepsysaurus pennsylvanicus.*

The available fish material is referrable entirely to *Dictyopyge macrurus* and *D. meekeri* (Schaeffer and McDonald, 1978). Although articulated skeletons occur in the lower and middle part of the member, only scattered fish scales and bones were recovered at the Doswell locality.

Branchiopods are locally common arthropods in the middle of the Falling Creek Member but none were seen at the Doswell locality.

A single cast of a shell from the lowest part of the Falling Creek Member demonstrates the presence of snails. In contrast, small unionid clams are abundant throughout the upper half of the formation, especially at the Doswell locality, where numerous clam casts were found scattered in the matrix from which came the holotype of *Doswellia.* Fecal pellets from pollen-

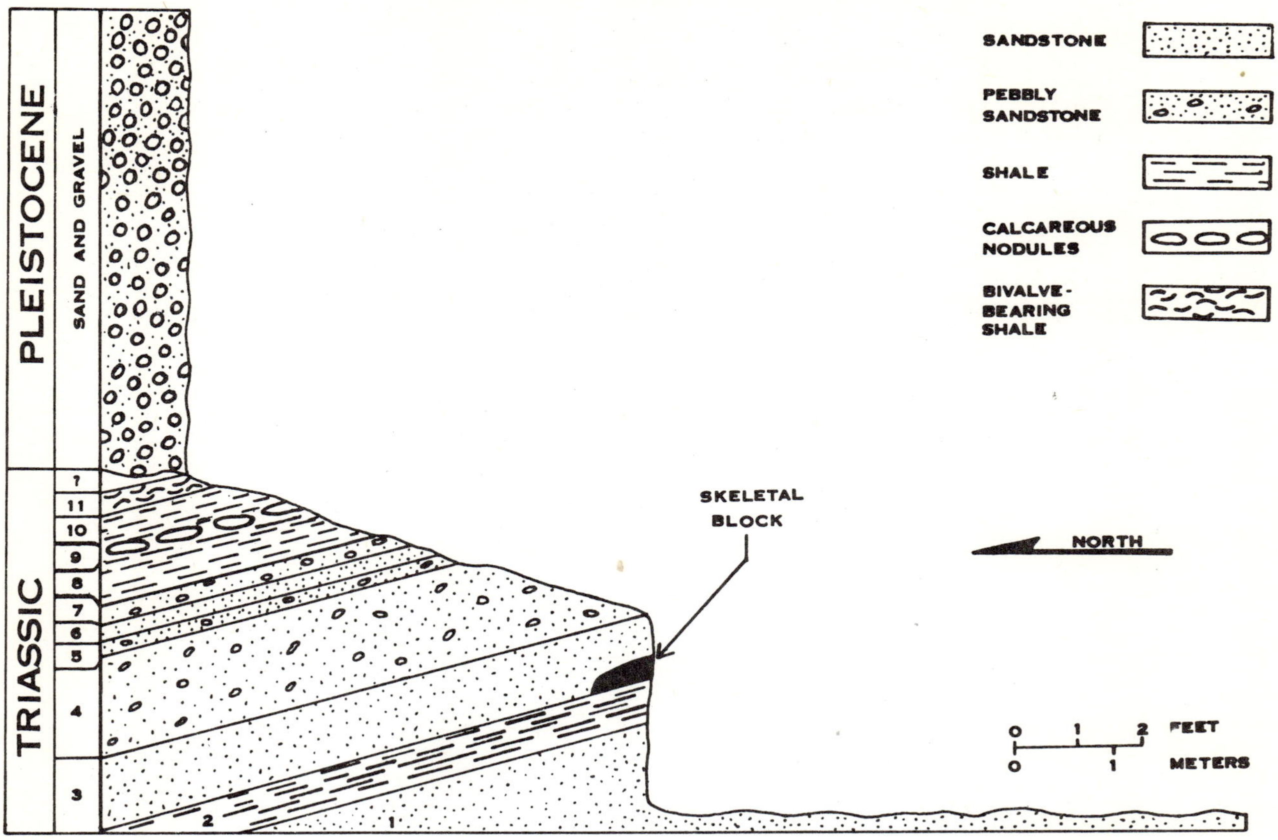

FIG. 5. Diagrammatic side view of the north wall of the Doswell sewer pit construction site showing relationships of the beds described in figure 5 and the location of the block in which the type skeleton of *Doswellia* was removed. The Triassic section is covered by 14 feet of Pleistocene terrace gravels. Vertical and horizontal scales are the same.

eating and algae-eating zooplankton record the presence of soft-bodied organisms at the Doswell locality not otherwise preserved (written communications: Bruce Cornet, 1976; Eleanora Robbins, 1977).

Abundant blue-green algal tissue is present in the algal-mat textured sediments at the Doswell site (Eleanora Robbins, written communication, 1977), while the immediately underlying shales yielded the following palynormphs (Bruce Cornet, written communication, 1976): *Alisporites aequalis, Aratrisporites scabratus, Colpectopollis ellipsoideus, Lagenella martinii, Lapposisporites cf. L. loricatus, Lunatisporites acutus, Ovalipollis ovalis, Patinasporites densus, Pityopollenites neomundanus, Protodiploxypinus doubingeri, Protohaploxypinus* sp., *Sulcatisporites kraeuseli, Triadispora modesta, T. verrucata, Vallasporites ignacii,* and *V. perspicuus.* In portions of the Falling Creek Member older than the Doswell locality the following additional palynomorphs have been identified; *Abiespollenites* sp., *Alisporites* sp., *Camerosporites secatus, Guttulapollenites* sp., *Klausipollenites* sp., *Ovalipollis ovalis, Patinasporites densus, Pityosporites* sp., and *Vallasporites ignacii* (Dunay and Fisher, 1974); *Alisporites ovatus, A. australis, Convolutiospora clausii, C. tuberculiforma, Neoraistrickia* sp., *Plicatisaccus badius,* and *Pseudoenzonalasporites summus* (Bruce Cornet, written communication, 1976). Although the palynoflora is quite varied, only unidentifiable scraps of macroscopic plant tissue were recovered at the Doswell locality. At other horizons of the Falling Creek Member, however, silicified log fragments referrable to the stem-genus *Auracarioxylon* are locally abundant (Knowlton, 1899). The palynoflora is similar to that of the Richmond basin (Bruce Cornet, 1977), so a macroflora similar to the one described from that basin by Fontaine (1883) should be expected.

TAPHONOMY

The type skeleton of *Doswellia* was semiarticulated (fig. 7) indicating it was partially disturbed after death. The posterior dorsal vertebral series was essentially intact and lying with the dorsal side up, as was the associated armor. The anterior dorsal vertebrae were scattered in an arc toward the animal's left, turning the neck and head region back toward the vicinity of the last presacrals. These vertebrae were found lying on their side. The anterior cervicals also lay on their sides, but were found in articulated position. The skull and jaws were displaced slightly to

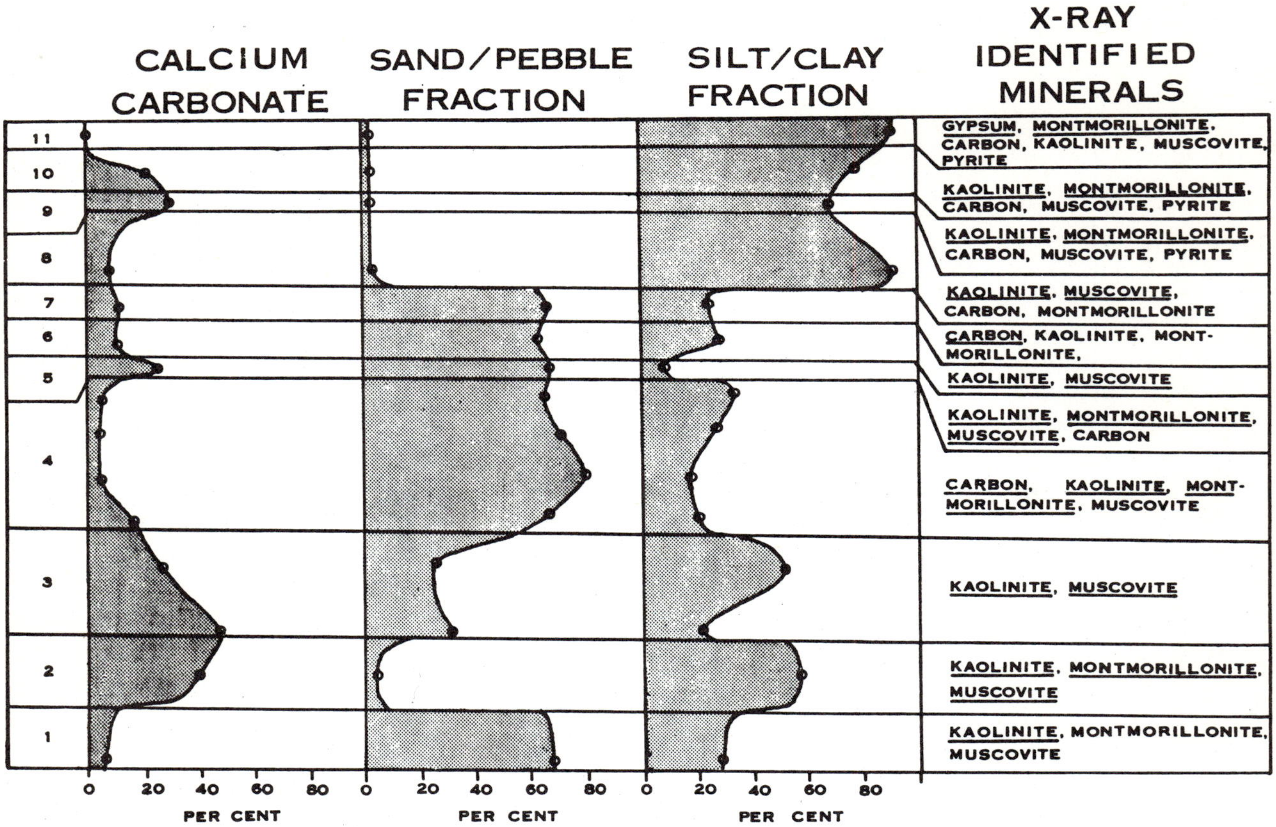

FIG. 6. Relative proportions of calcite, sand/pebble and silt/clay fractions of the beds described in figure 5. The carbonate fraction was determined by placing samples in hydrochloric acid and the resultant disaggregated material was sieved to determine the coarser and finer than sand fractions. Materials listed in the clay and silt fraction were determined by X-ray diffraction analysis. Underlined minerals are species which produced the highest peaks in each run.

the left of the cervical vertebrae. The entire sacral complex was displaced to the right of the posterior, dorsal vertebrae and then jumbled after displacement. The anterior caudals were found with the sacral complex, but all of the more posterior caudals had been scattered in a direction roughly parallel with the axis of the posterior dorsal series. Since one pubis was punctured by a round hole identical in size and shape to several anterior parasuchian teeth found in the Doswell pit (pl. 7, figs. D, F), it seems likely that the scattering observed in the skeleton was the result of parasuchian scavenging. Lack of any extensive damage, however, suggests that the carcass was abandoned because it was too rotten to be desirable by the time it was discovered and disturbed.

Although the skull was slightly crushed, most of the bone is remarkably well preserved and uncrushed. This is probably due to early cementation of the rock by the abundant calcite.

SYSTEMATICS

CLASS: Reptilia
SUBCLASS: Archosauria
ORDER: Thecodontia
SUBORDER: Doswelliina, new suborder
FAMILY: Doswelliidae, new family

Doswellia, gen. nov.

Type species: *Doswellia kaltenbachi* gen. et sp. nov.

Distribution: Upper Triassic (Middle Carnian), Falling Creek Member, Doswell Formation; Taylorsville Basin, Virginia (USA)

Generic Diagnosis: Body armored, at least dorsally and laterally (minimum of 10 longitudinal columns of armor in the posterior dorsal and sacral region), plates imbricated except for a nuchal plate behind the head and a region of suturally articulated armor in the sacral region; body elongated and flattened, crocodile-like; neck elongate; head about 15 cm (6 inches) in length, flattened with narrow elongate rostrum; tail flattened dorso-ventrally, wide at base but tapering rapidly posteriorly. Total length about 2 meters (6 feet).

Skull sculptured, akinetic except for minor kinesis of the quadrates and pterygoids, containing a single pair of upper temporal fenestrae, which are located at the convergence of the parietal, postorbital and squamosal bones ("euryapsid" configuration); orbits close together and directed upward; jugals expanded and

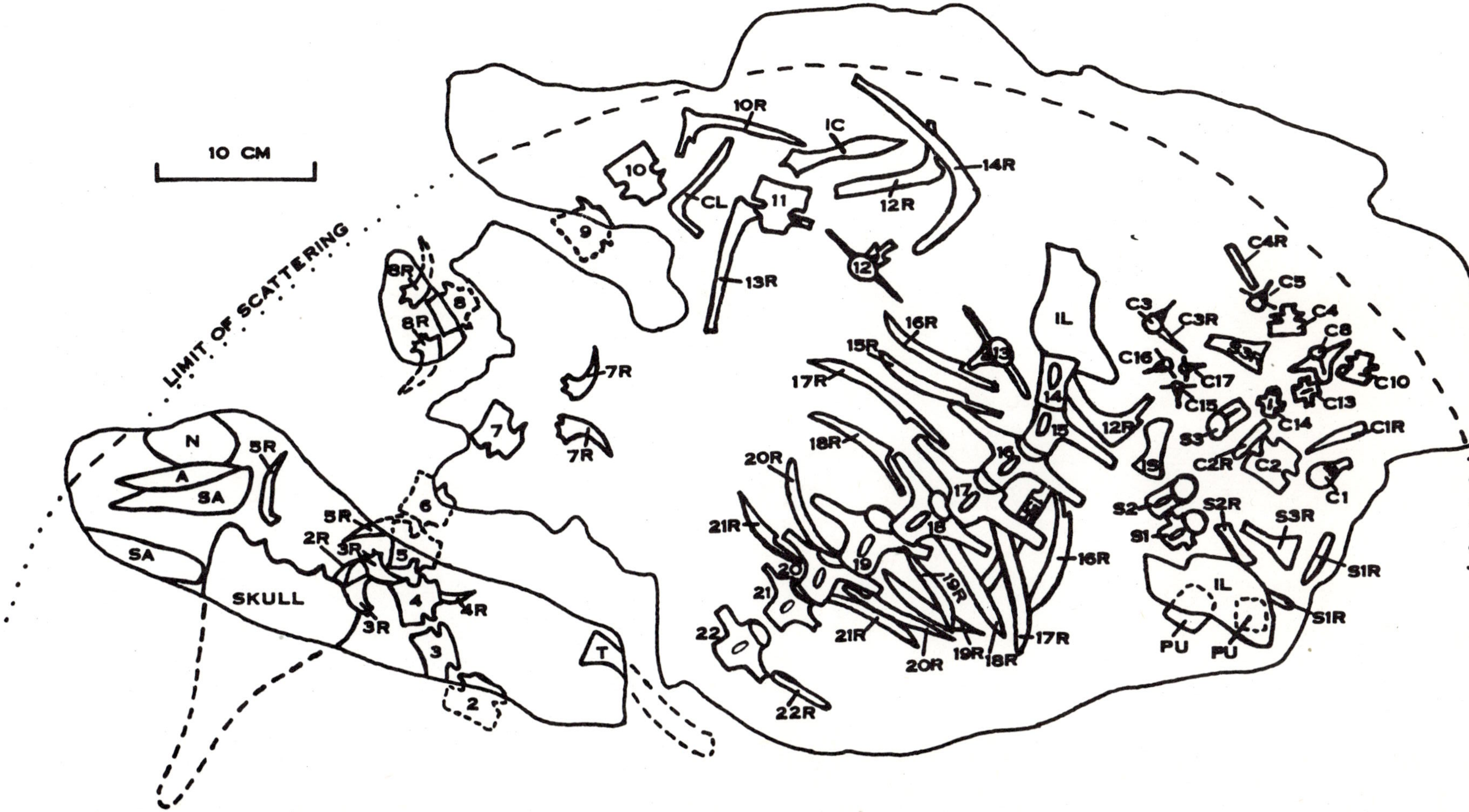

FIG. 7. Distribution of bones in the blocks removed from the Doswell sewer plant. Position of the two small blocks containing portions of the neck and skull is inferential. SK—skull, A—angular, SA—surangular, presacral vertebrae numbered 2-22, S—sacrael vertebrae, C—caudal vertebrae, R—rib of respectively numbered vertebra, I—interclavicle, CL—clavicle, IL—ilium, PU—pubis, IS—ischium, T—tibia, N—nuchal collar. Other portions of armor are not shown.

enlarged, postfrontals and postparietals not present, squamosals form a pair of prominent, posteriorly directed "horns," supraoccipital crest present; interpterygoid vacuity closed, pterygoid flanges wide and thin; small, round, blunt pterygoid teeth present in three rows; suborbital vacuity present at convergence of the pterygoid, ectopterygoid, and maxillary elements; basipterygoid processes extend down to roof of mouth and are surrounded anteriorly, laterally, and posteriorly by the pterygoids; ventral plate of epipterygoid ossified but dorsal portion apparently cartilaginous; no laterosphenoid or other braincase ossification anterior to basisphenoid; no post-temporal fenestrae; quadrate with prominent otic notch, anteriorly directed process extends along dorsal surface of quadrate ramus of pterygoid; stapes "sauropsid"; exoccipitals and opisthotics fused. Mandible deep posteriorly, articular and surangular both contribute to articular facet; prominent retroarticular processes; dentaries anteriorly fused into symphysis; lateral mandibular fenestra absent; external jaw surface sculptured. Teeth thecodont and homodont; probably recurved, conical and fluted.

Vertebrae amphicoelous; cervicals elongate and lacking a ventral keel; posterior dorsals with elongate transverse processes which bear a ventral groove in which the rib neck fits; three sacral vertebrae, one incorporated from the dorsal series; anterior caudals bear ribs, posterior bear either fused ribs or transverse processes; neural spines low throughout vertebral series. Presacral ribs dichocephalous, including first sacral incorporated from dorsal series; posterior sacrals and caudals holocephalous; cervical ribs and anterior six dorsal ribs are sharply angled, all more posterior ribs are straight; third dorsal through first sacral ribs attach only to transverse process of respective vertebra; first caudal rib pair arcuate along anterior border to fit around the posterior border of the ilium.

Clavicle sharply angled; interclavicle cross-shaped and flattened. Anterior process of the ilium sharply curved laterally above acetabulum and enlarged, lateral margin fluted; plate-like pubis and ischium downturned distally but not greatly elongated; small fenestra developed between pubis and ischium, obturator foramen separate and not incorporated into fenestra; acetabulum round, deep, large, open ventrally but with a prominent ridge developed dorsally. Femur sigmoidal and apparently with a cartilaginous head, internal trochanter present, fourth trochanter absent.

Doswellia kaltenbachi sp. nov.

Derivation of name: *Doswellia* for the town of Doswell near which the type skeleton was discovered; *kaltenbachi* for James Kaltenbach who discovered the type specimen.

Holotype: USNM 244214; axial skeleton from seventh cervical through fifth caudal, scattered more posterior caudals, associated ribs; pelvis complete except for left ischium; clavicle, interclavicle; dorsal and lateral armor badly scattered except for an articulated patch from the posterior dorsal region. USNM 214823 from the same locality and bedding plane consists of the postorbital $\frac{1}{3}$ of the skull, posterior $\frac{1}{3}$ of the jaws, second through fifth cervicals, cervical ribs, nuchal plate, cervical armor plates, and distal end of a (?)tibia; probably of the same individual as USNM 244214, but since the association is not absolutely certain, to avoid confusion it is numbered separately and designated a paratype. Discovered in pit dug for foundations at the Doswell sewer plant located one-fourth mile northwest of the confluence of the North Anna and Little Rivers, near Doswell, Hanover County, Virginia. Upper Falling Creek Member, Doswell Formation (Taylorsville Basin), Newark Supergroup. USNM 244214 collected by James Kaltenbach, Dr. Nicholas Hotton III, and Robert E. Weems, September, 1974. USNM 214823 collected by Donald Beazely, September, 1974.

Specific diagnosis: same as for genus.

Distribution: same as for genus.

Referred specimens: (1) USNM 186989; one cervical vertebra, posterior dorsal vertebra, left dentary, left femur, scattered armor plates. Northeast streambank of creek on east side of Richmond, Fredericksburg, and Potomac Railway 1.2 miles north of Ashland, Hanover County, Virginia. Upper Falling Creek Member, Doswell Formation, (Taylorsville Basin), Newark Supergroup. Collected by Robert E. Weems and Dr. J. B. Weems, March, 1968. (2) USNM 244215; one anterior dorsal vertebral centrum, same locality as (1); collected by David H. Massie and Shelton P. Applegate, no date (probably late 1950's).

SKULL

The skull of the paratype of *Doswellia* is not preserved anterior to the rear of the orbits. A single isolated jugal element (pl. 1, figs. A, B), recovered from the Doswell site, confirms the shape and anterior extent of that element. The dentary found at the Ashland site (USNM 186989), if correctly assigned, indicates the shape of the snout, but otherwise features of the anterior skull are unknown. It is entirely possible that the front of the snout was downturned as in parasuchians and *Proterosuchus,* but in the absence of positive evidence this has not been assumed. No secondary palate is indicated in the material at hand.

The preserved part of the skull roof is flat, slightly crushed dorso-ventrally and is covered with a vermiform sculpture similar to but in detail not identical with that seen on crocodilian and labyrinthodont skulls. Orbits and temporal fenestrae are dorsal and close to the midline (fig. 8). There is no evidence of a pineal foramen, but a prominent fracture which runs

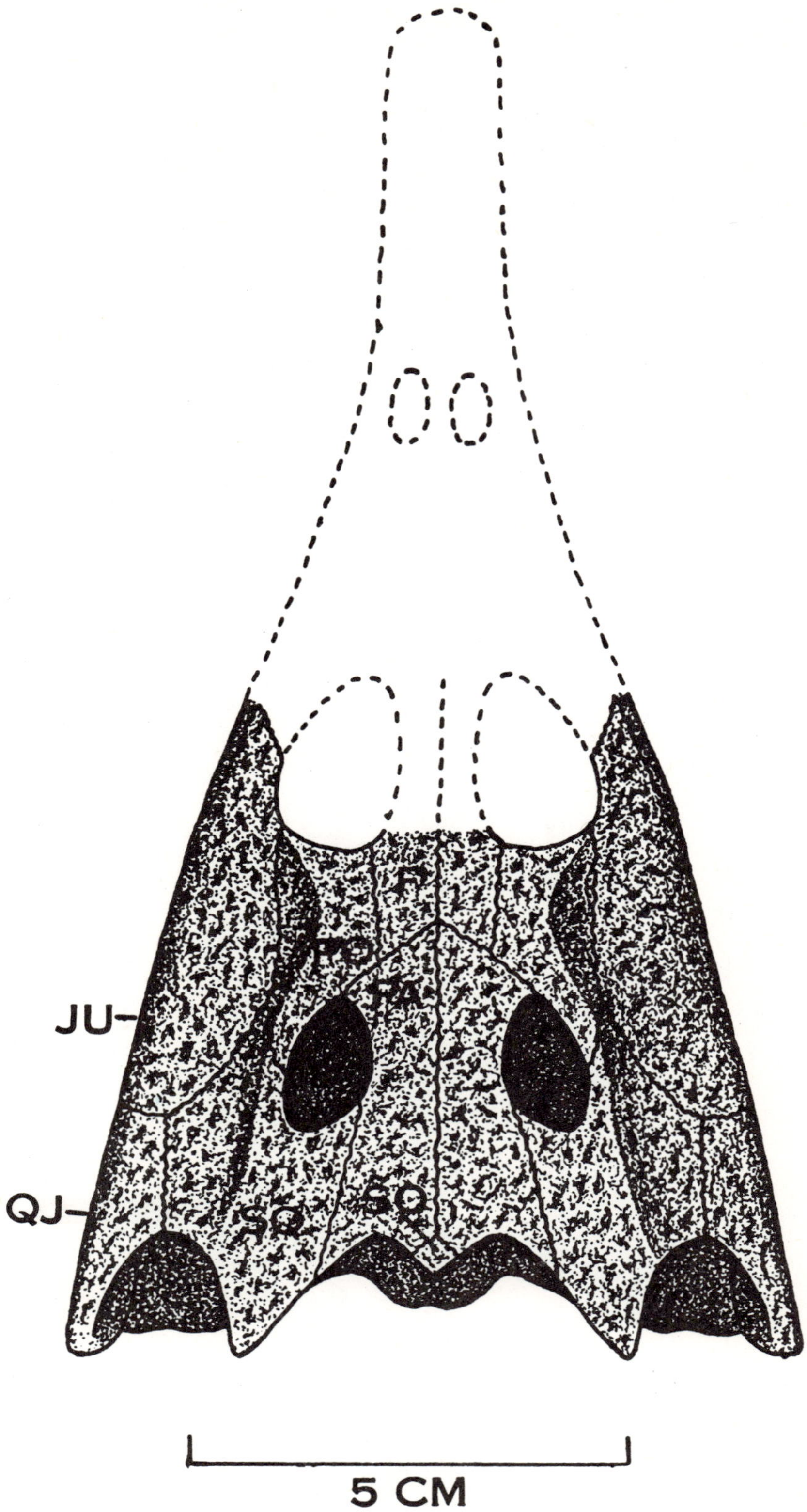

FIG. 8. Dorsal view of the skull of *Doswellia kaltenbachi*. F—frontal, JU—jugal, PA—parietal, PO—postorbital, QJ—quadratojugal, QU—quadrate, SO—supraoccipital, SQ—squamosal.

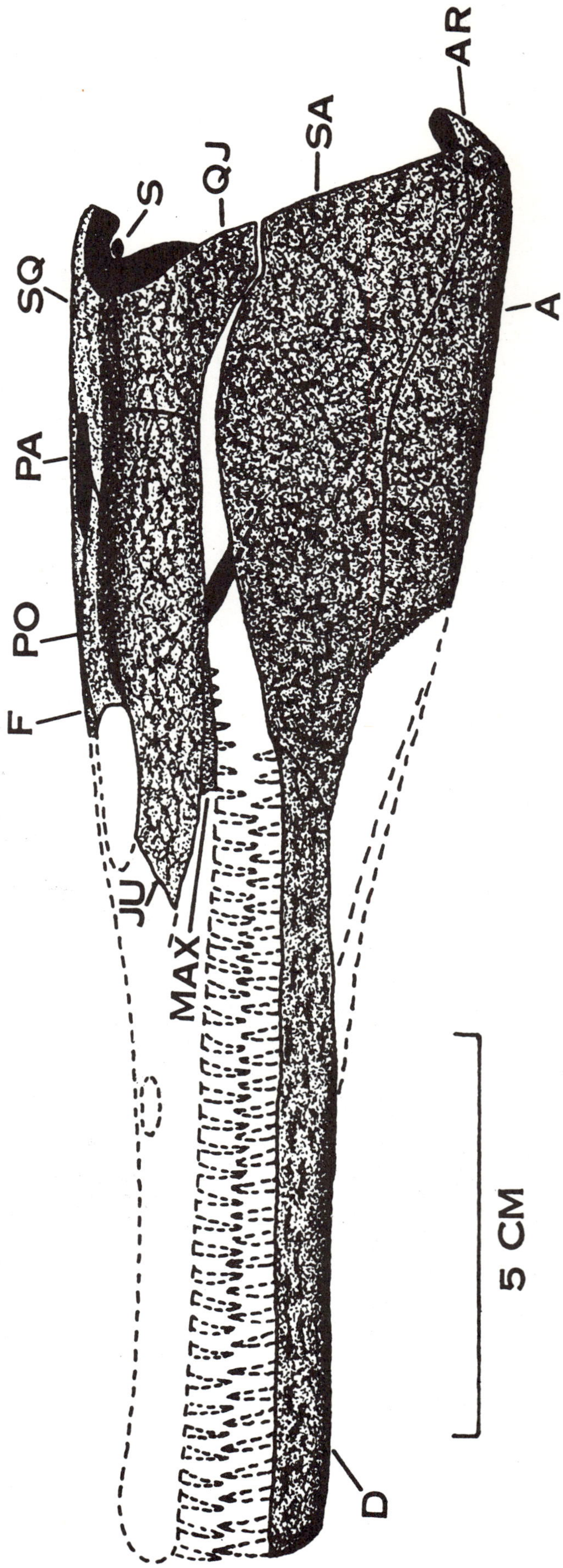

FIG. 9. Lateral view of the skull and jaws of *Doswellia kaltenbachi*. A—angular, AR—articular, D—dentary, F—frontal, JU—jugal, MAX—maxilla, PA—parietal, PO—postorbital, QJ—quadratojugal, S—stapes, SQ—squamosal, SA—surangular.

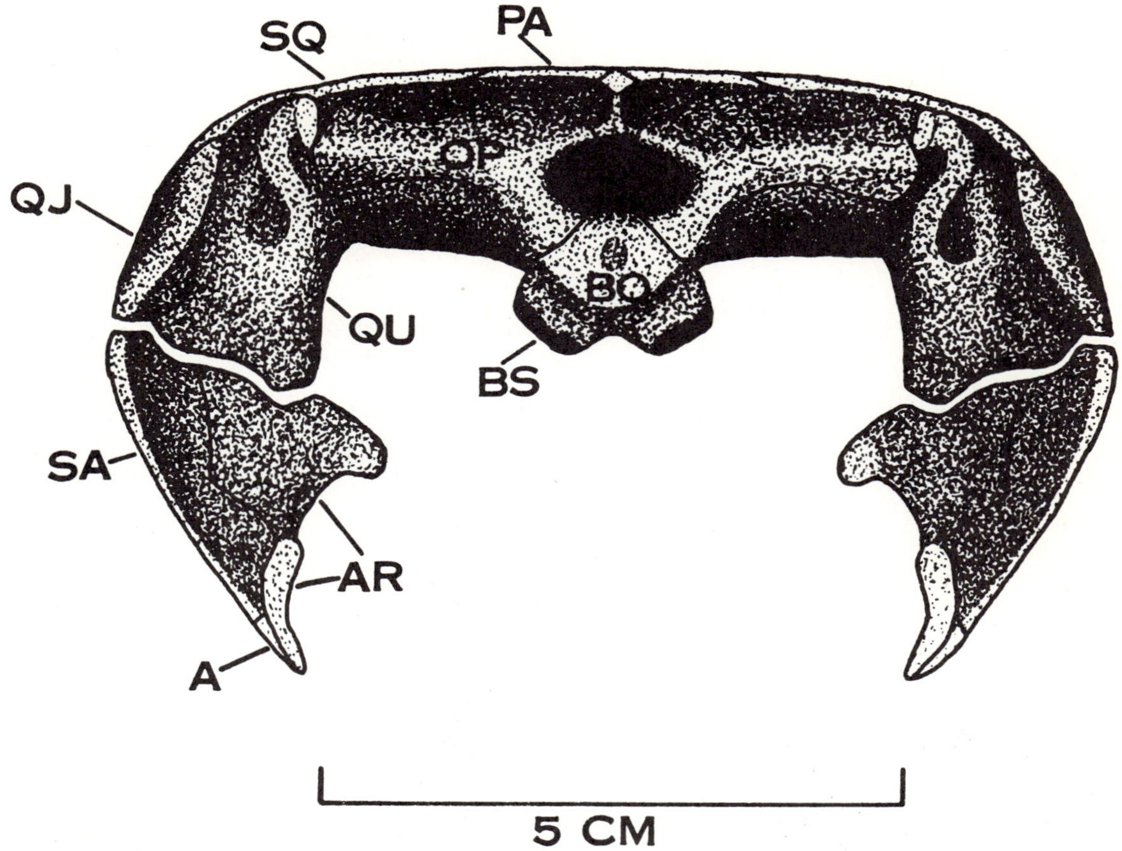

FIG. 10. Occipital view of the skull and jaws of *Doswellia kaltenbachi*. A—angular, AR—articular, BO—basioccipital, BS—basisphenoid, OP—opisthotic and exoccipital, fused, PA—parietal, QJ— quadratojugal, QU—quadrate, SA—surangular, SQ—squamosal, SO—supraoccipital.

transversely across the parietals could conceivable have obliterated all trace of a very small foramen if it had been present. Frontals and parietals are very narrow and the postfrontal has disappeared, apparently as a consequence of the constriction of space between the orbits. Correlated with this narrowing of the medial elements is a major expansion of the cheek region, mostly by enlargement of the jugal (fig. 9). Postorbital and squamosal elements are present; these, along with the parietals, bound a single pair of temporal fenestrae in the "euryapsid" position. The jugal, postorbital, quadratojugal and squamosal are tightly sutured together and afford no hint that any other opening was ever present. The quadratojugal forms part of the quadrate-articular joint in primitive fashion (e.g. *Proterochampsa* (Sill, 1967)) (fig. 10).

The occipital margin of the skull is indented on either side of the midline, presenting the shape of an "E" in dorsal aspect (fig. 8). The unsculptured supraoccipital projects backward from the parietals as a rather turtle-like median crest, forming the central bar of the "E," and two prominent, anomalous squamosal "horns" project to the rear at the postero-

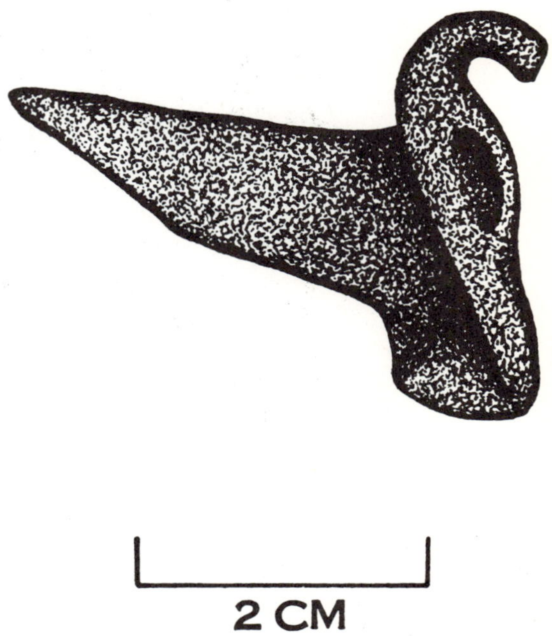

FIG. 11. External lateral view of quadrate of *Doswellia kaltenbachi*. Note anteriorly directed pterygoid process and rounded border of the cephalic condyle.

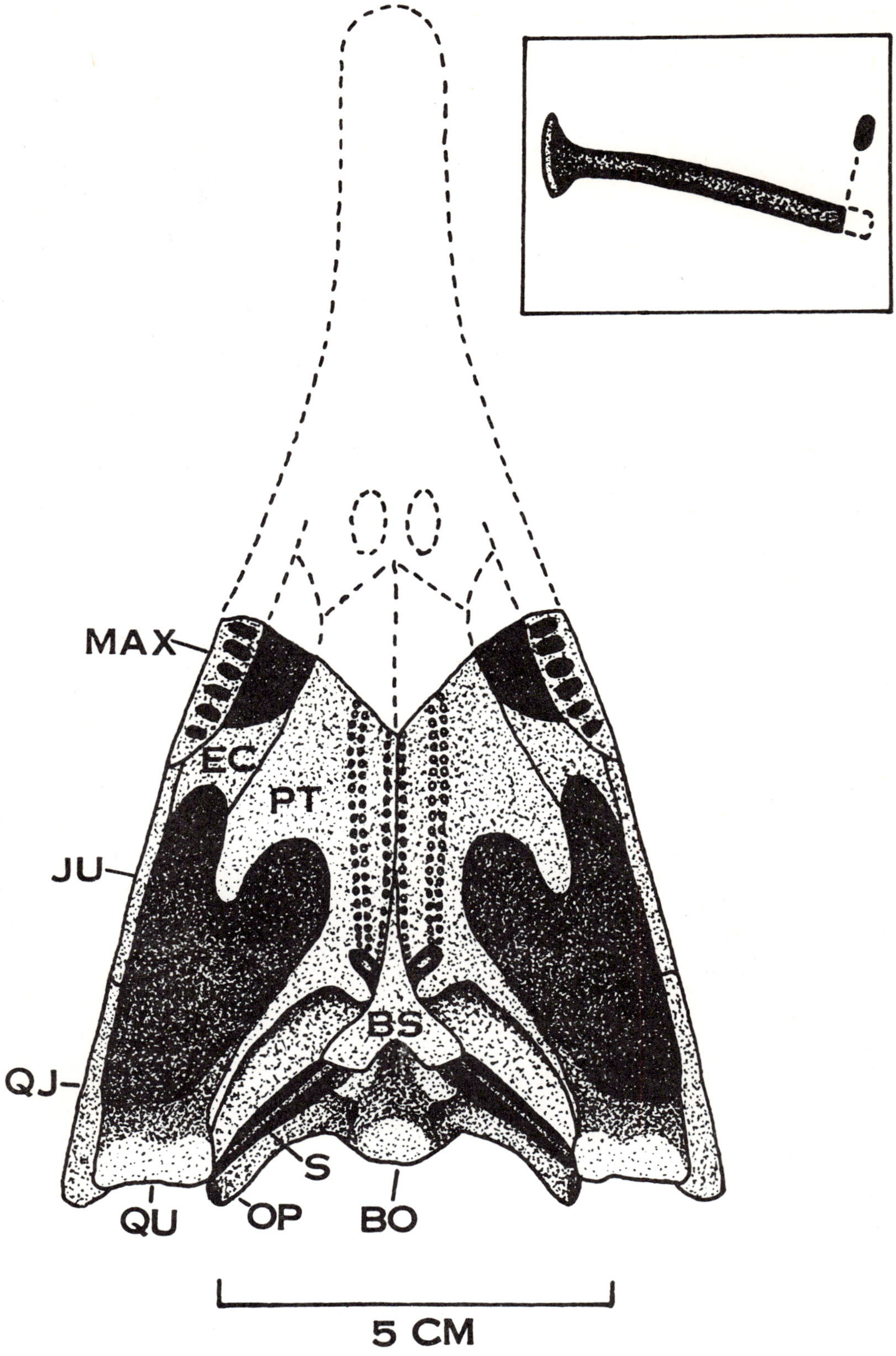

FIG. 12. Ventral view of the skull of *Doswellia kaltenbachi*. BO—basioccipital, BS—basisphenoid, EC—ectopterygoid, JU—jugal, MAX—maxilla, OP—opisthotic and exoccipital, fused, PT—pterygoid, QJ—quadratojugal, QU—quadrate, S—stapes. Inset: Stapes of *Doswellia kaltenbachi* shown at twice the scale of the ventral skull view.

lateral corners of the skull roof to complete the pattern.

The occipital region is low but otherwise of a normal sauropsid build. Post-temporal fenestrae are absent. The quadrates are wide and massive and together with the quadratojugal form a nonplanar locked articular surface which would have allowed no lateral motion at the joint. When viewed laterally, the quadrates lie in a nearly vertical plane due to a forward migration of the quadrate head (fig. 11). The highly curved proximal end of each quadrate is strongly directed posteriorly, forming a prominent tympanic notch (fig. 8).

On the ventral surface of the skull (fig. 12), only the posterior end of the maxillaries are preserved, which include six empty alveoli. The posterior palate is broadly floored by the pterygoids which are conjoined and abut at a pair of planar joints with the basisphenoid, obliterating the interpterygoid vacuity. Alveoli, some containing small round, blunt teeth, define the location of three rows of small teeth on each pterygoid which apparently ran forward beyond the portion of the pterygoids that are preserved. Two of these rows are located close to each other near the medial border of the pterygoid, but the third row is widely separated from the other two and lies close to the midline next to the basisphenoid. Posteriorly all three rows converge toward the basipterygoid processes. These processes are encircled laterally, anteriorly, and posteriorly by the pterygoids, but not medially. Posteriorly from the basipterygoid processes the pterygoids extend posterolaterally and lie below anteriorly directed pterygoid processes of the quadrates. The ectopterygoids are sutured to the pterygoids and maxillaries, but are bounded anteriorly by a smooth beveled border which must be the posterior rim of a suborbital vacuity lying between the maxillary, ectopterygoid, and palatine elements. Posteriorly the ectopterygoids and pterygoids together form a flattened and wide posteriorly protruding shelf which extends into the subtemporal fossa. This shelf is dorsoventrally thinner than, but otherwise similar to, the ectopterygoid-pterygoid flange found in crocodilians and is unlike the more knob-like thickened flange found in lizards. The ventral portion of the epipterygoid is ossified and broadly sutured to the pterygoid which partially wraps around the base of the brain cavity. No trace of the dorsal ascending process of the epipterygoid was found, however, so probably it was cartilaginous or absent. There is no evidence that there was an osseous laterosphenoid element or any other element bounding the anterior brain wall in front of the basisphenoid. The dorsum sellae extends upward at its lateral margins to form two prominent knobs which produce a partial "C"-shaped encirclement of the trigeminal nerve.

The stapes lie within a well-developed otic notch in the quadrates. Although much thicker than the comparable elements in lizards and crocodiles, these bones are comparable in size and shape to the stapes of most Triassic archosaurs and are much longer and thiner than the stapes of synapsids, ichthyosaurs, or *Araeoscelis*. The footplate is elliptical in shape, being wider than high, and there is no trace of an arterial foramen (fig. 12). Thus the stapes of *Doswellia* have a slender "sauropsid" configuration and not a massive "theropsid" configuration.

Measurements of selected skull parameters and elements are shown in table 2.

TABLE 2

SELECTED MEASUREMENTS FROM THE PRESERVED PORTION OF THE SKULL OF *Doswellia kaltenbachi* (in mm.)

Measurement	mm.
Width as preserved	84.6
Length as preserved	66.0
Total depth (skull roof to quadrate base)	28.4
Depth from skull roof to base of basisphenoid	24.1
Minimum distance between postorbital fenestrae	13.5
Midline length of parietals	31.6
Midline length of supraoccipital on skull roof	8.6
Distance between distal ends of opisthotics	48.1
Maximum length of quadratojugal	15.1
Distance between distal ends of the posterior borders of the parietals	25.8
Distance between posterior tips of the squamosal "horns"	47.4
Height of the basioccipital condyle	8.1
Width of the basioccipital condyle	9.3
Height of the foramen magnum	5.1
Width of the foramen magnum	14.5
Height of the supraoccipital (skull roof to foramen magnum)	6.4
Maximum width of the quadrate condyle	15.4

MANDIBLE

A dentary was found at the Ashland site (USNM 186989) and can be attributed to *Doswellia*. Posterior jaw elements are known with the paratype, and together these various elements permit a restoration of the low jaw in lateral, dorsal, and posterior aspects (figs. 9, 10, and 13). The jaws are sculptured on their external surface with a pattern much like that on the skull and meet anteriorly along a broad symphysis between the dentaries producing a "Y" shape when seen in dorsal aspect. Apparently there was no external oval fenestra between the dentary, surangular and angular elements.

The dentary contains 33 alveoli which are rounded and indicate the former presence of a thecodont dentition. The first two alveoli are crowded together and are somewhat smaller than the third alveolus but from the third alveolus posteriorly the sockets are evenly spaced at intervals less than the alveolar diameter. The alveoli become progessively smaller toward the rear of the mouth. The anterior tip of the dentary is somewhat enlarged into a bulbous shape, and a flattened and grooved surface is present along

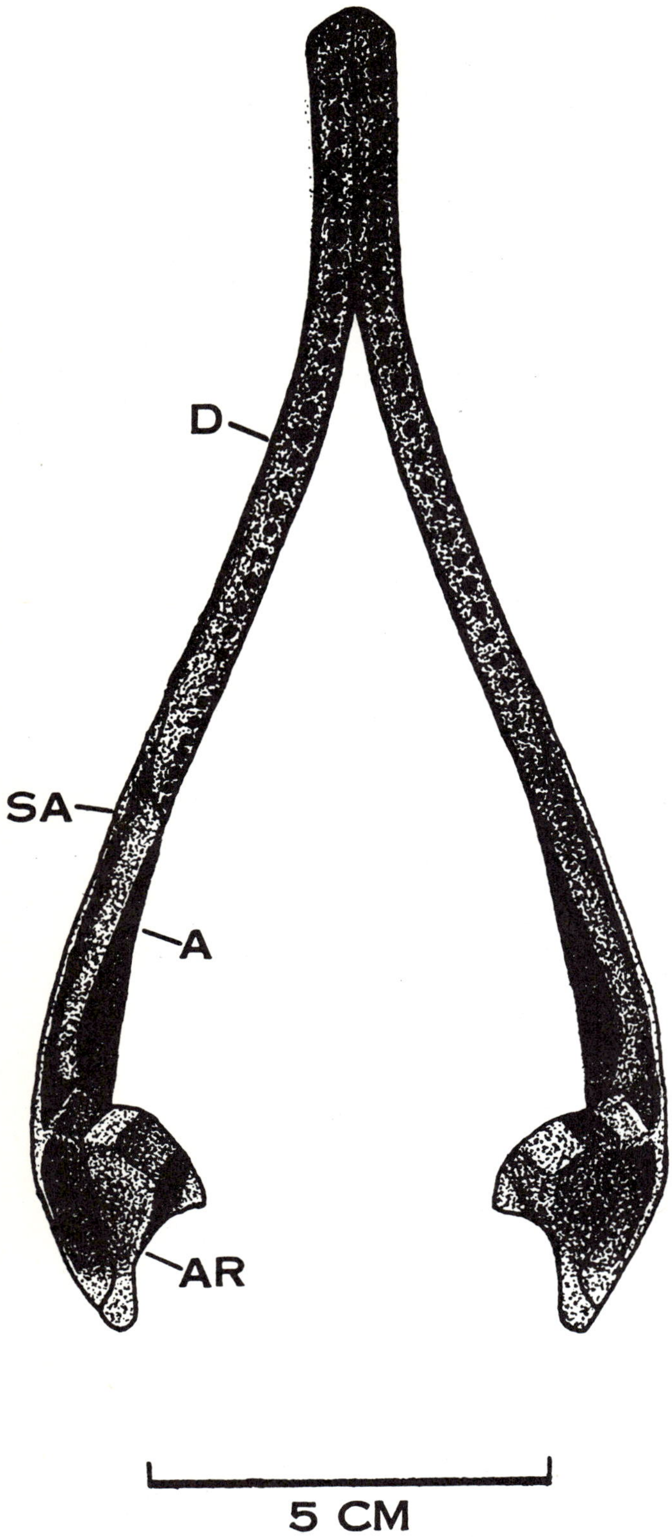

FIG. 13. Dorsal view of the low jaws of *Doswellia kaltenbachi*. A—angular, AR—articular, D—dentary, SA—surangular.

the anteromedial one-third of its length where the two dentaries met in a symphysis. It is not known whether the splenial formed part of the external jaw surface but this is a common condition in many reptiles. In lateral aspect, the angular forms an elongate plate along the posteroventral margin of the jaw, covering the external surface of the retroarticular process. The surangular forms a much greater portion of the jaw ramus than the angular and is convexly curved along its dorsal margin, filling a dorsally concave area formed by the ventral borders of the jugal and quadratojugal. However, this is not a true coronoid process; no separate coronoid element was present. The articular is compact dorsally (fig. 13), forming a wide but socket-like articulation contact with the quadrate. This would have permitted no sliding motion of the jaw hinge in any direction. The external one-fourth of the articulation surface is not formed by the articular, but rather by the surangular. The retroarticular process lies far below the level of the glenoid and was the site of insertion for a very long and powerful depressor mandibulae muscle. The entire posterior jaw ramus and retroarticular region are extremely deep and must have housed a massive set of adductor and extensor musculature.

TEETH

No marginal teeth were found in the alveoli of either the paratype skull of *Doswellia* or in the associated dentary from the Ashland specimen. This indicates that the connective tissues holding the teeth in their sockets were not at all calcified. However, a number of small teeth were found scattered throughout the matrix at the Doswell site which are of the right size and basal shape to belong to *Doswellia,* so are tentatively referred to this species. The teeth are small (less than one cm. in total length), conical, recurved and fluted, with rounded bases (pl. 7, fig. E). They are of a shape that is generally associated with piscivorous animals whether reptiles, mammals, or amphibians. The fluting is reminiscent of that seen on the much larger teeth of *Rutiodon carolinensis* Emmons (1856) and the slightly larger teeth of *"Rutiodon" sulcatus* Emmons (1857).

A few teeth are still present in the alveoli of the pterygoids. They are low, round knobs similar in size and shape to the type of teeth developed in pycnodont fishes and similar in shape to the teeth of placodonts but smaller. Such teeth would seem to be designed to crush food rather than pierce it as would the presumed marginal teeth.

VERTEBRAE

The vertebral column of *Doswellia* varies considerably along its length and discrete cervical, dorsal, sacral, and caudal regions can be distinguished. Measurements of the preserved portions of all the vertebrae are shown in tables 2–5. Figure 14 shows diagrammatically the parameters measured.

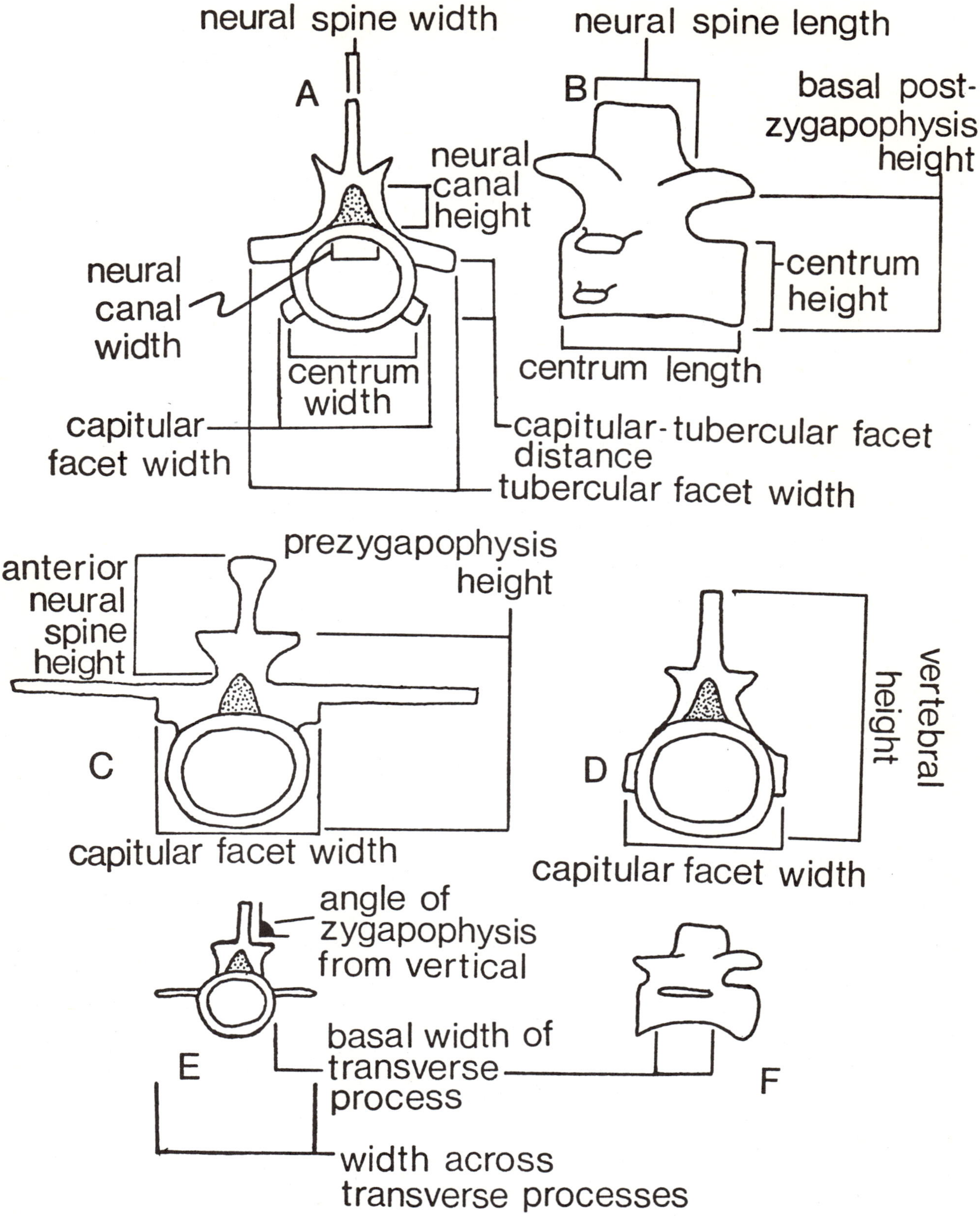

FIG. 14. Diagram showing parameters measured on vertebrae of *Doswellia kaltenbachi* in the cervical vertebrae (A, B) and (where different) in the dorsal (C), posterior sacral and anterior caudal (D), and posterior caudal (E, F) vertebrae.

TABLE 3

MEASUREMENTS OF CERVICAL VERTEBRAE OF *Doswellia kaltenbachi* (in mm.)

	Centrum					Neural canal				Distance between	
	Anterior		Posterior			Anterior		Posterior			
Number	Height	Width	Height	Width	Length	Height	Width	Height	Width	Capitular facets	Tubercular facets
2	—	—	—	—	—	—	—	—	—	—	—
3	12	14	13	15	31	5	8	5	8	17	22
4	13	15	14	15	32	5	8	6	7	18	23
5	14	16	—	—	—	6	8	—	—	19	23
6	—	—	—	—	—	—	—	—	—	—	—
7	15	18	16	18	32	6	10	6	8	$20^{\pm}$	26
8	15	18	16	18	31	7	11	6	11	21	27

	Distance between capitular and tubercular facets	Zygapophyses						Neural spine		
			Facet			Facet		Anterior		
Number		Height to top	Length	Width	Height to base	Length	Width	Height	Width	Length
2	—	—	—	—	—	7	6	5^{+}	1	25
3	2	26	8	7	18	8	6	8	2	13^{+}
4	3	27	10	7	21	8	7	6^{+}	$2^{\pm}$	22
5	4	25	8	7	—	—	—	7^{+}	2	26^{+}
6	—	—	—	—	—	8	7	9^{+}	2	—
7	4	26	6^{+}	7	$22^{\pm}$	10	9	16	2	15
8	4	25	11	8	21	—	—	—	—	—

Number	Total height of vertebra	Angle of pre-zygapophyses from vertical	Angle of post-zygapophyses from vertical
2	—	—	40°
3	24^{+}	40°	40°
4	28^{+}	40°	40°
5	26^{+}	40°	—
6	—	—	40°
7	36	40°	55°
8	—	65°	—

Cervical Vertebrae: The cervical region was badly damaged when the skeleton was unearthed so only vertebrae 2–5 can be conclusively placed in their proper positions. In this series the anterior central faces seem to become regularly wider and higher. The next cervical behind the fifth which has a well preserved anterior face seems to belong two places farther back in the progression and would therefore be the seventh cervical. A damaged, neural arch and spine in front of the well preserved putative seventh cervical is therefore considered to be a remnant of the sixth cervical vertebra. Thus, probably no cervicals are entirely missing except the first, though it is still possible that one or more were present in the portion of matrix now missing between the block containing the trunk and skull. On available evidence, the best presacral count is 22 (8 cervicals and 14 dorsals), upon which further references to vertebrae and rib numbers are based.

The axis is recognizable because of its enlarged neural spine, the only part of this vertebra which is preserved. Relative to the cervicals immediately behind it, the axial spine is long and has a strongly developed, anteriorly directed protuberance for muscle attachment to the supraoccipital-parietal crest at the back of the skull. The remaining cervicals (3–8) all have elongate centra of about equal length. The height and width of each increases posteriorly, however, so the posterior cervicals are proportionately more robust than the anterior ones. The central faces are round and amphicoelous except for the eighth, which has an anterior face noticeably wider than high. There is no trace of intercentra or of ventral keels on the centra. Parapophyses for double-headed cervical ribs are present on all of these vertebrae and the distance between the capitular and tubercular facets remains about the same (3–5 mm) throughout the sequence. The neural canals become higher and wider posteriorly through the sequence. The prezygapophyses are longer than the postzygapophyses and the articular facets of both sets are longer than wide. In most, these facets are angled at *ca.* 40° from vertical; in the eighth, the angle is lowered to 65° from vertical, forming a transition to the still lower, dorsal series of zygapophyses. The neural spines are progressively longer along the cervical series, being almost non-

TABLE 4

MEASUREMENTS OF DORSAL VERTEBRAE OF *Doswellia kaltenbachi* (in mm.)

	Centrum					Neural canal				Distance between	
	Anterior		Posterior			Anterior		Posterior			
Number	Height	Width	Height	Width	Length	Height	Width	Height	Width	Capitular facets	Tubercular facets
1	—	—	—	—	—	—	—	—	—	—	—
2	15	20	15	20	30	5	11	6	11	24	47+
3	15	20	14	19	29	—	10	—	11	25	58
4	15	20	14	19	30	6	9	5	—	23	65
5	15	18+	14	18	29	4	7	—	8	22	68±
6	14	19	14	19	29	4	8	4	8	25	84±
7	14	19	14	19	30	5	8	4	7	26	97
8	14	19	14	20	30	5	9	—	—	28	100
9	14	19	14	19	30	5	9	—	9	25	100
10	14	19	14	19	29	5	9	3	10	25	89
11	14	19	14	20	28	4	8	3	8	25	81
12	14	20	14	21	24+	5	8	5	8	24	71
13	14	20	14	20	26	6	9	5	10	23	47+
14	—	20	—	—	—	5	10	—	8	22	50+

	Zygapophyses						Neural spine			
		Facet			Facet		Anterior			
Number	Height to top	Length	Width	Height to base	Length	Width	Height	Width	Length	Total height of vertebra
1	—	—	—	—	9	9	—	5	14	—
2	24	9	8	21	8	7	15	6	15	37
3	—	—	—	—	—	—	—	7	16	36
4	—	—	—	18	6	6	15	6	17	36
5	—	—	—	—	—	—	11+	6	16	36
6	19	—	—	18	—	—	12	6	16	35
7	19	5	6	16	5	4+	13	6	16	32
8	20	6	6	16	—	—	10+	6	19	31
9	21	7	6	17	6	4	15+	6	16	34±
10	21	6	6	16	6	5	12	5	16	33
11	21	7	6	16	7	6	13	6	15	34
12	22	7	6	19	7	7	13	6	19	33
13	23	7	5	21	9	8	12	5	17	34
14	—	7	7	—	8	8	12	6	16	—

Number	Angle of pre-zygapophyses from vertical	Angle of post-zygapophyses from vertical
1	—	70°
2	65°	70°
3	—	—
4	—	80°
5	—	—
6	90°	90°
7	90°	85°
8	90°	85°
9	90°	85°
10	90°	85°
11	90°	85°
12	85°	75°
13	85°	75°
14	65°	70°

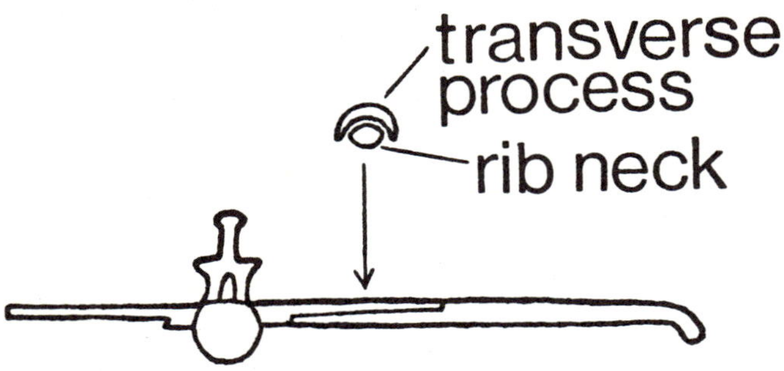

FIG. 15. Diagram showing articulation system of the dorsal ribs and vertebrae of *Doswellia kaltenbachi.*

existent on the third cervical but quite prominent by the seventh and eighth. All cervical neural spines are thin and bladelike, in contrast to the dorsal series where the dorsal tips are swollen into bulbous knobs.

Dorsal vertebrae: The dorsal centra decrease slightly in length posteriorly throughout the series. The central faces are amphicoelous, wider than high, and of comparable dimensions throughout the series. In the dorsal series the tubercular facets rapidly expand out to form elongate transverse processes. These lengthen back to the seventh and eighth dorsals and then shorten back to the fourteenth. The distance between the capitular facets increases only slightly but the capitular facets move up and back so that by the fifth dorsal the capitular facet is located directly beneath the transverse process. Concomitantly, the edges of the transverse processes are curved downward to form a long groove beneath the transverse process between the capitular and tubercular facets. The neck of the rib was thus snugly confined in this groove (fig. 15). Such a system of rib-vertebra articulation is unknown in any other animal. The neural canals are low and wide, remaining nearly constant in dimensions. Neither prezygapophyses nor postzygapophyses are greatly elongated. The angles of the facets have flattened further from the angle seen in the cervicals to about 75° from vertical. The neural spines are long and low, each having a distinct bulbous knob on its dorsal tip.

Sacral vertebrae: The 23rd vertebra in the original presacral series has its posterior central border suturally joined to the 24th vertebra but otherwise it bears a rib and transverse processes comparable to those found on the posterior dorsals. This vertebra has obviously been incorporated into the sacral complex from the presacral series. The second sacral vertebra is suturally joined by its centrum to the vertebra in front and behind it while the third is suturally joined only anteriorly. The second and third sacrals (and first caudal) are the only vertebrae in which the neural arches are only suturally attached to the respective centra, indicating adult size had almost but not quite been reached. The second and third sacrals have no transverse processes; the holocephalous sacral ribs attach directly and broadly to the centra. The neural spines on all of these vertebrae are low and bear swollen knobs at their distal tips as in the dorsal series.

Caudal vertebrae: Five anterior caudals were found along with portions of their respective ribs. When reassembled on the assumption that the vertebrae are consecutive, the ribs are regularly shorter to the rear, so as to produce a lateral narrowing of the tail which matches the curvature of at least one piece of caudal armor. The inference that these vertebrae represent a continuous series therefore appears reasonable. The caudals bear facets for chevrons starting with the second. All of the vertebrae become somewhat smaller and shorter posteriorly. No transverse processes are present, but caudal ribs attach to the anteromedial portion of each centrum directly. The neural spines are low and by the fourth have lost the distal knob which is so prominent on the dorsal and sacral vertebrae. The articular facets of the postzygapophyses lie in a nearly horizontal plane, while the articular facets of the prezygapophyses are "L"-shaped (see caudal sections in fig. 25). This produces a strongly locked articulation between these vertebrae.

Behind the fifth caudal at least three vertebrae are missing. The next two, presumed to be the eighth and tenth caudals, have long and wide blade-like processes extending laterally from the centrum. These are either transverse processes or more probably caudal ribs which have entirely fused to their respective centra. The eighth caudal is quite unusual in that the anterior face of the centrum is strongly angled posteroventrally (pl. 4, fig. M). This must have produced a pronounced downward flexure of the tail and indicates that the tail in front of this point must have been carried essentially parallel to the ground. Behind this point more vertebrae are missing and the exact location of all more posterior caudals is speculative. Probably they are close to caudals 13, 17–20, and 25. In the region of caudals (?)13–(?)20 the lateral processes, though still blade-like, become steadily shorter and have their distal ends less expanded. The centra in this region have become elongate and spool-like. A small posterior caudal, possibly caudal 25, is from well behind the others. It is much smaller than the other caudals and the lateral processes and neural arch have shortened even further than in the preceding caudals. Otherwise it is simply a smaller version of the ones located anteriorly to it.

CHEVRONS

All chevrons associated with the skeleton of the type specimen, even those associated with the anterior caudals, are short, never exceeding the height or proportions of the neural arches and spines of the

TABLE 5

MEASUREMENTS OF SACRAL AND ANTERIOR CAUDAL VERTEBRAE OF *Doswellia kaltenbachi* (in mm.) (N/A = not applicable)

	Centrum					Neural canal				Distance between	
	Anterior		Posterior			Anterior		Posterior			
Number	Height	Width	Height	Width	Length	Height	Width	Height	Width	Capitular facets	Tubercular facets
Sacral											
1	15	22	$13^{\pm}$	19	26	11	8	5	12	25	38
2	14	18	13	20	27	6	8	6	9	24	N/A
3	11	20	$12^{\pm}$	19	28	3	11	3	8	25	N/A
Caudal											
1	13	18	11	18	26	8	4	12	7	21	N/A
2	12	20	12	16	24	6	7	—	—	19	N/A
3	12	18	12	15	24	$4^{\pm}$	$7^{\pm}$	—	7	18	N/A
4	12	15	13	13	22	5	9	—	7	17	N/A
5	11	13	14	12	23	5	7	—	6	15	N/A

	Capitular facet		Zygapophyses						Neural spine		
				Facet			Facet		Anterior		
Number	Height	Width	Height to top	Length	Width	Height to base	Length	Width	Height	Width	Length
Sacral											
1	5	7	25	10	6	17	6	7	15	6	15
2	15	12	22	6	3	19	6	4	13	5	17
3	10	15	18	6	4	$16^{\pm}$	8	5	10^{+}	5	18
Caudal											
1	14	15	23	8	4	22	8	4	9	5	19
2	11	16	21	8	4	21	—	—	9	6	17
3	9	15	—	—	—	—	—	—	—	—	—
4	8	15	20	—	—	$21^{\pm}$	—	—	$8^{\pm}$	4	14
5	7	14	19	—	—	—	—	—	$8^{\pm}$	3	14

Number	Total height of vertebra	Angle of pre-zygapophyses from vertical	Angle of post-zygapophyses from vertical
Sacral			
1	35	60°	80°
2	34	70°	60°
3	25^{+}	55°	90°
Caudal			
1	32	20°	90°
2	29	20°	—
3	—	—	—
4	30	25°	—
5	29	—	—

caudal region. Therefore, it seems likely that none of the chevrons were greatly elongate. The best preserved chevron is shown in plate 4, figure O. The length of the chevrons and neural spines suggests that the tail was wide and flat anteriorly, becoming nearly round posteriorly after about the tenth caudal due to a rapid decrease in the length of the caudal ribs.

RIBS

The ribs of *Doswellia kaltenbachi* are differentiated according to their position: cervical, dorsal, sacral, and caudal. In addition quite different anterior and posterior dorsal ribs can be distinguished as well.

Cervical Ribs: The cervical ribs are less well preserved than those in other parts of the body. All that were recovered are dichocephalous (double-headed) with extremely short necks. Capitular and tubercular facets are only 3–5 mm apart. The ribs are laterally directed in their proximal 20–30 mm, and then curve sharply to the rear. At the point of curvature a keel is developed which gives the ribs a plow-shaped appearance. Preserved portions of these ribs, except the two attached to the second and third cervicals, have thin but very wide shafts which give the ribs a very robust appearance. All are broken distally, however, so long and thin shafts may have extended pos-

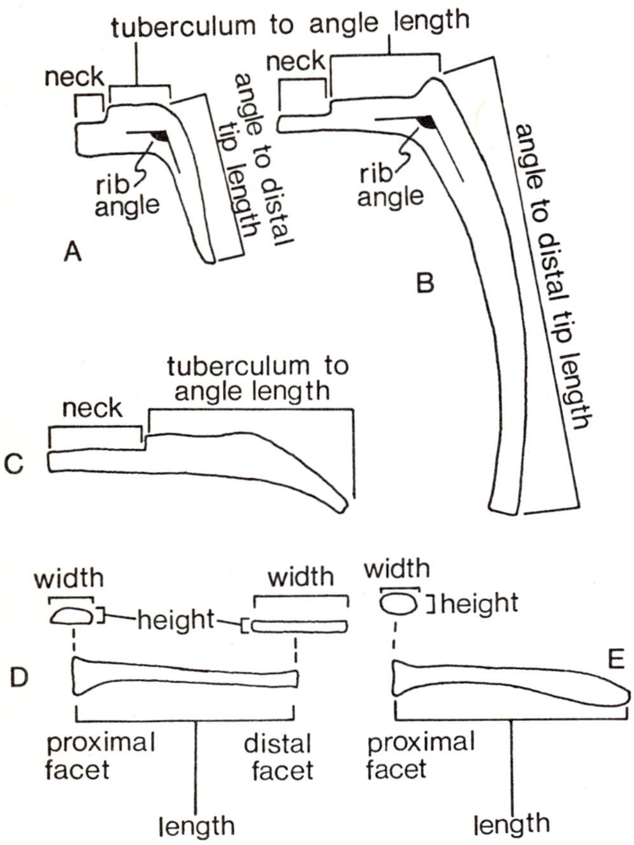

FIG. 16. Diagram showing parameters measured on ribs of *Doswellia kaltenbachi* in the cervical (A), anterior dorsal (B), posterior dorsal (C), sacral (D), and anterior caudal (E) region.

teriorly from the preserved portions to give a slightly more normal appearance. The external border of the posteriorly directed distal shaft of each rib has a large, flattened, rugose surface presumably for armor attachment.

Anterior Dorsal Ribs: The anterior dorsal ribs are directed sharply posteroventrally in their distal two-thirds to three-fourths, presumably with respect to their relationship to the pectoral girdle and its musculature. In this configuration they contrast sharply with the entirely laterally directed, posterior dorsal ribs. The first anterior dorsal rib shows a marked thinning of the laterally directed, proximal shaft. The distal portion is not preserved. In its proportions this fragmentary rib seems transitional between typical cervical and dorsal ribs. The second through fifth dorsal ribs are essentially complete. They show a progressive posterior migration of the capitular attachment surface so that the rib neck ultimately comes to be in contact with the transverse process of the vertebra with which it is attached. Concomitantly a progressive lengthening of the proximal, laterally directed rib shaft and a progressive lengthening of the distal, posteroventrally directed shaft occur. By the third dorsal rib there is a pronounced dorsally directed protuberance at the angle of the rib shaft which seems to represent an enlarged version of the keels developed in the comparable region of the cervical ribs. This protuberance is even more pronounced on the fourth dorsal rib but is progressively less well developed on the fifth and sixth dorsals. The dorso-external margin of the second dorsal rib bears a rugose area for armor attachment much reduced in size from the scars seen in the posterior cervical ribs. These scars are even smaller in dorsal ribs three, four, and five. No rib behind five bears this rugose scar. The sixth dorsal rib has a prominent angle as do the ribs before it but in the sixth rib the shaft terminates immediately below the flexure at an articular surface. The element to which this surface attached was not found and may well have been cartilaginous. In any event, some structure continued posteroventrally from this articular surface and for this reason it is included as an anterior dorsal rib. In other respects, however, it is more like the more posterior dorsal ribs and thus is transitional between the two regions.

Posterior Dorsal Ribs: The posterior dorsal ribs, seven through fourteen, are directed laterally and lack ventral flexure in the rib shaft, dorsally directed keels or knobs and rugose surfaces on the latero-external border for armor attachment. All have the capitulum, neck, and tuberculum nestled tightly against the ventral surface of the respective, transverse process. Seven through ten have distal ends tapering to a point, though by ten the point is rather blunt. Eleven through fourteen have essentially rounded, distal ends. This sequence of ribs becomes progressively shorter posteriorly, both in the length of the neck and in the length of the shaft.

Sacral Ribs: The fifteenth rib in the dorsal series is similar to the fourteenth but slightly longer. However, since its distal end is in contact with the anterior border of the ilium and its associated vertebra is suturally joined to the vertebra behind it, the fifteenth rib appears to be functionally a part of the sacrum. The shaft is still small in diameter relative to its length, as in the preceding dorsal ribs. The second sacral rib, by contrast, has a markedly expanded distal end for attachment to the ilium and a stout and thickened shaft. Its holocephalous (single-headed) proximal end is attached suturally to the anterior portion of the centrum of the second sacral vertebra and to the posterior portion of the centrum of the first sacral vertebra. The third sacral has comparable proximal and distal attachment points but differs in that the distal end of the rib is even more expanded antero-posteriorly to provide a strong attachment most of the length of the ilium.

TABLE 6

MEASUREMENTS OF POSTERIOR CAUDAL VERTEBRAE OF *Doswellia kaltenbachi* (in mm.)

Number	Centrum					Neural canal				Width across transverse processes	Basal width of transverse processes
	Anterior		Posterior			Anterior		Posterior			
	Height	Width	Height	Width	Length	Height	Width	Height	Width		
8(?)	—	12	13	13	$22^{\pm}$	5	6	6	4	52^{+}	13
10(?)	—	—	11	11	—	—	—	—	4	$48^{\pm}$	11
13(?)	11	11	12	$11^{\pm}$	23	5	4	4	4	$30^{\pm}$	11
14(?)	12	11	—	—	—	4	6	—	—	$24^{\pm}$	$9^{\pm}$
15(?)	10	10	12	9	23	4	5	3	4	24	8
16(?)	10	9	10	10	22	—	4	—	—	22	11
17(?)	—	—	10	9	22	—	—	—	—	22	7
25(?)	7	5	8	7	20	—	—	—	3	16	8

Number	Zygapophyses						Neural spine			Total height of vertebra
	Height to top	Facet		Height to base	Facet		Anterior		Length	
		Length	Width		Length	Width	Height	Width		
8(?)	—	—	—	19	6	4	11	2	$10^{\pm}$	$26^{\pm}$
10(?)	—	—	—	—	—	—	—	—	—	—
13(?)	19	6	4	—	—	—	—	—	13	21^{+}
14(?)	18	5	3	—	—	—	—	—	—	20^{+}
15(?)	17	—	—	—	—	—	8	1	12	22
16(?)	14	—	—	—	—	—	—	1	—	19^{+}
17(?)	—	—	—	13	5	2	—	1	—	20^{+}
25(?)	—	—	—	—	—	—	—	—	—	—

Number	Angle of prezygapophyses from vertical	Angle of postzygapophyses from vertical
8(?)	—	40°
10(?)	—	—
13(?)	20°	—
14(?)	20°	—
15(?)	15°	—
16(?)	20°	—
17(?)	—	85°
25(?)	—	—

Caudal Ribs: The anterior caudal region bears discrete single-headed ribs broadly and rugosely attached only to the anterior border of their respective vertebrae. Of the five vertebrae preserved from the anterior caudal region, all bear separate caudal ribs. Beyond this point at least three vertebrae and ribs are missing, and the next two vertebrae bear either totally fused ribs or transverse processes. Thus separate ribs can be demonstrated no further back than the fifth. The first caudal rib is distally subrounded in cross-section and is curved distinctly anteriorly. This shape causes its distal end to terminate at the posterior border of the ilium in a manner analogous to that of the first sacral rib. Since, however, the rib head does not attach to the third sacral centrum and the first caudal vertebra is not suturally joined to the third sacral vertebra, this rib is treated as a first caudal rib and not a fourth sacral rib even though it appears to be partly incorporated functionally into the sacrum. The second and fourth caudal ribs are well preserved and are progressively shorter in their shaft length and become increasingly flattened dorsoventrally at their distal ends. The proximal ends become smaller but their basic shape and mode of attachment to the vertebrae remains unchanged.

At least sixteen ventral rib cage elements (pl. 3, fig. H) were found representing eight or more pairs of bony elements arrayed at the ends of possibly a majority of the dorsal ribs. There was a smaller but comparable element found in articulated position at the distal end of the second dorsal rib, so presumably the large elements begin behind this point and extend back at least as far as the tenth dorsal vertebra. Alternatively, two pairs may have been present back only to the sixth vertebra.

TABLE 7

MEASUREMENTS OF CERVICAL, DORSAL, AND THE FIRST SACRAL RIBS OF *Doswellia kaltenbachi* (in mm.) (*N/A* = not applicable)

Number	Left or right	Neck length	Tuberculum to angle length	Angle to distal length	Angle of rib shaft flexure
Cervical					
2	left	—	13+	29+	—
3	right	2	18+	19+	—
4	right	2	21	29+	120°
5	right	2	22	39+	110°
6	—	—	—	—	—
7	left	2	24	34+	110°
8	right	2	30	44+	110°
Dorsal					
1	both	4	21+	—	—
2	right	9	27	88	110°
3	left	16	30	112	100°
4	left	20	38	110	90°
5	both	26	46	109	100°
6	right	31	45	17	110°
7	both	38	57	0	N/A
8	both	35	59	0	N/A
9	right	31	61	0	N/A
10	right	31	56	0	N/A
11	right	24	50	0	N/A
12	right	23	44	0	N/A
13	both	19	41	0	N/A
14	right	15	41	0	N/A
Sacral					
1	right	7	45	0	N/A

GASTRALIA

In addition to the massive elements mentioned previously, numerous thin and elongate pieces of bone apparently representing more ventromedial elements of the rib cage and possibly abdominal ribs were scattered through the matrix (pl. 3, fig. G). None of these are well preserved, but they are almost certainly similar to the ventromedial gastralial elements found in living lizards and crocodilians.

PECTORAL GIRDLE

Only two elements of the pectoral girdle, the interclavicle and one clavicle, are certainly identifiable among the remains of *Doswellia.*

Interclavicle: The interclavicle (pl. 5, fig. C; fig. 17) is long and robust, being much like the interclavicle of pelycosaurs and cotylosaurs except that it is not so upturned at its anterior end. A pair of large facets are developed along either side of the midline on the anterior border for articulation to the clavicles but no more posteriorly located facets are developed for attachment with the coracoid.

Clavicle: One clavicle, presumed to be the right, is

TABLE 8

MEASUREMENTS OF POSTERIOR SACRAL AND ANTERIOR CAUDAL RIBS OF *Doswellia kaltenbachi* (in mm.) (N/A = not applicable)

Number	Left or right	Length	Proximal facet		Distal facet	
			Height	Width	Height	Width
Sacral						
2	left	41	11	16	14	17
3	right	62	11	17	14	43±
Caudal						
1	right	83	9	15	N/A	N/A
2	right	79	11	16	N/A	N/A
3	right	27+	9	14	N/A	N/A
4	right	62	9	13	N/A	N/A
5	right	—	7	7	N/A	N/A

present (pl. 3, fig. F). The clavicle runs posterolaterally and then laterally from the interclavicle finally turning sharply upward and posteriorly at about a 90° angle flexure (fig. 25). The bone is flattened laterally where it runs upward and is rather flattened dorsoventrally elsewhere along its length. It is thickest about half way along its total length.

PELVIC GIRDLE

The pelvic girdle is well preserved and has been fully restored (figs. 16–18) in dorsal, ventral, lateral and anterior aspects.

Ilium: The most prominent feature of the ilium is the strongly lateral flexion of its dorsal margin. This blade is quite long and is sculptured along its lateral edge. A small forward projection affords attachment for the end of the first sacral rib. Along the internal border just at the level of the lateral flexure is a prominent scar for the reception of the ends of the second and third sacral ribs. Articular surfaces for the pubis and ischium are roughly planar and directed anterodorsally and posterodorsally respectively; in detail these surfaces are rugose.

Pubis: The pubis is plate-like and rather primitive in construction except for being slightly downturned anteriorly. The obturator foramen is an anteroposteriorly elongate hole entirely surrounded by the pubis. A rugose surface is developed anteriorly, presumably for attachment of the rectus abdominis musculature. Posteriorly the pubis thins to a beveled edge, and apparently there was a small fenestra between the pubis and ischium, filled in life by cartilage. Medially the pubes meet along a mutually sutured symphesis.

Ischium: The ischium is covered with small areas of spongy anomalous bone growth, perhaps a consequence of inflammation. The anterior border tapers to a beveled edge while medially the two ischia must have met suturally. The posterior border of the ischium is strongly downturned and bears a large and prominent scar posterolaterally, presumably for attachment of the musculature retracting the femur. Though strongly downturned, the rear of the ischium is not much elongated.

Acetabulum: The acetabulum is imperforate but large and deep. About three fifths of it is made by the ilium and about one fifth each by the pubis and ischium. Ventrally the acetabulum is open, but anteriorly, dorsally, and posteriorly it is bordered by a rim which is sharpest and most prominent directly above the top of the acetabulum.

LIMBS

Only the femur and the distal end of a tibia may be associated with the skeleton of *Doswellia*. The former was recovered with the Ashland specimen and the latter with the paratype from Doswell.

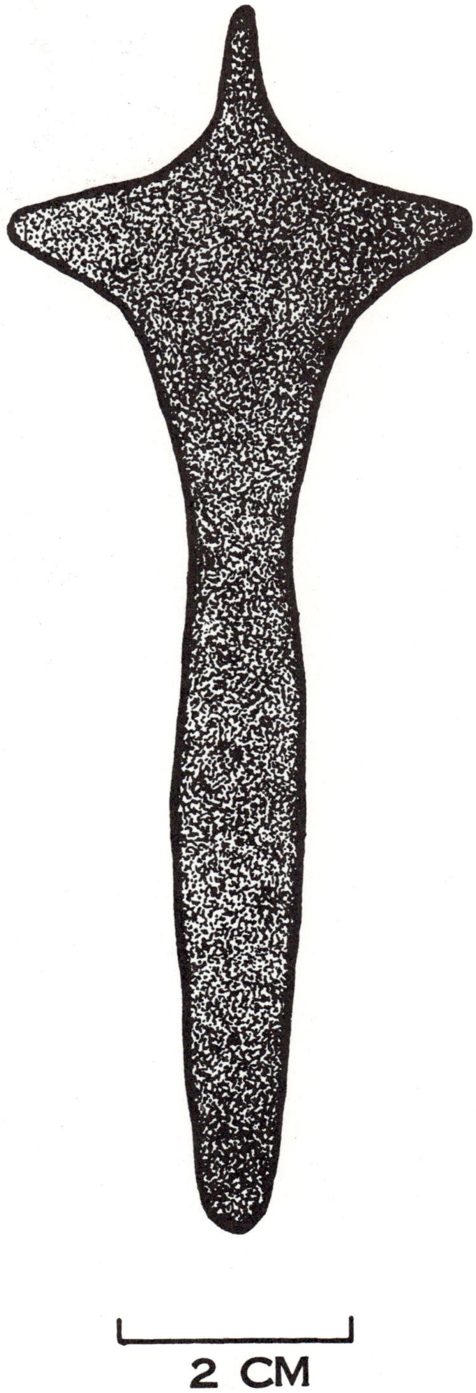

FIG. 17. Interclavicle of *Doswellia kaltenbachi*.

Femur: The femur (pl. 5, figs. A, B; fig. 22) is sigmoidal in shape, being quite similar in gross shape to the femur of parasuchians and crocodiles. There is no indication of a fourth trochanter, but an internal trochanter is present. The proximal end is rugose, which suggests the presence in life of a terminal car-

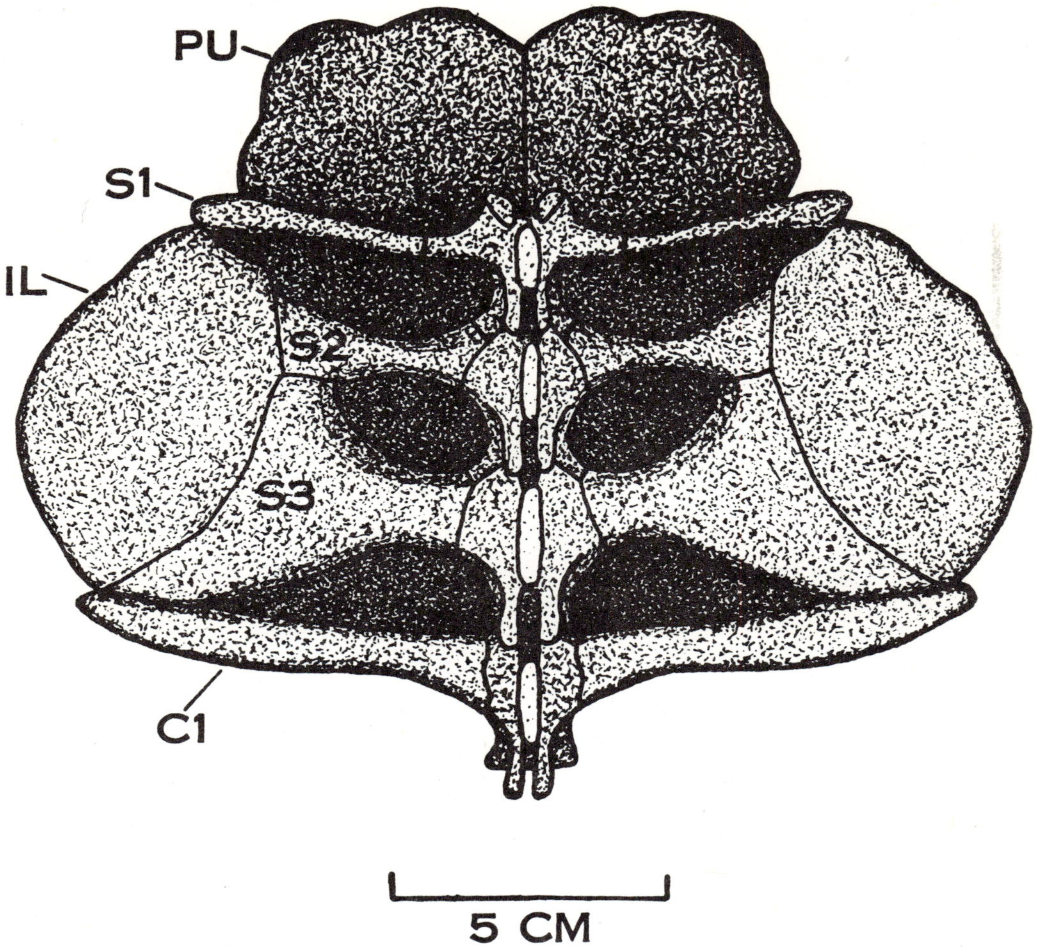

FIG. 18. Dorsal view of the pelvis of *Doswellia kaltenbachi*. C—caudal rib, IL—ilium, PU—pubis, S—sacral vertebra.

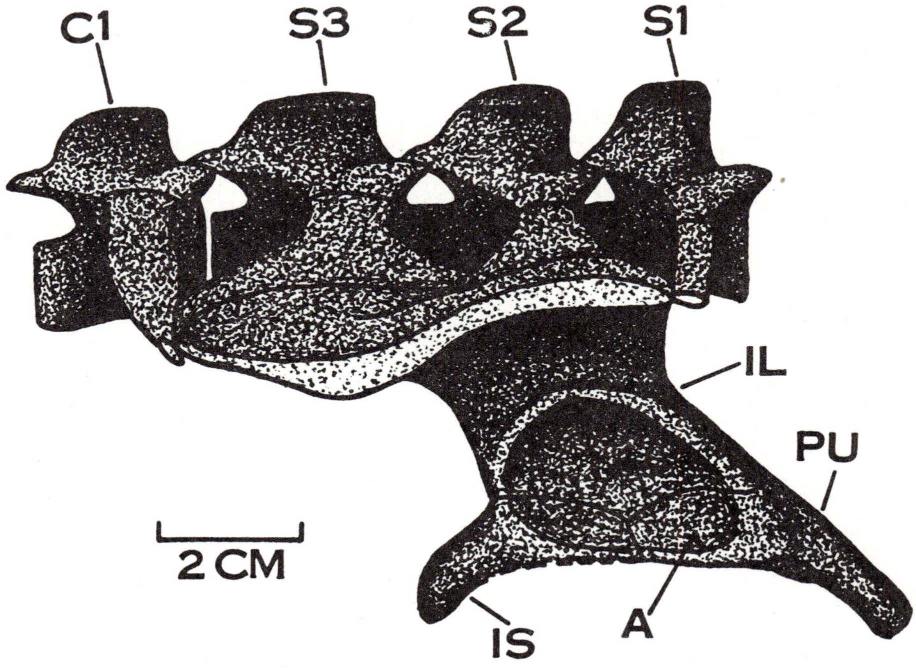

FIG. 19. Lateral view of the pelvis of *Doswellia kaltenbachi*. A—acetabulum, C—caudal vertebra, IL—ilium, IS—ischium, PU—pubis, S—sacral vertebra.

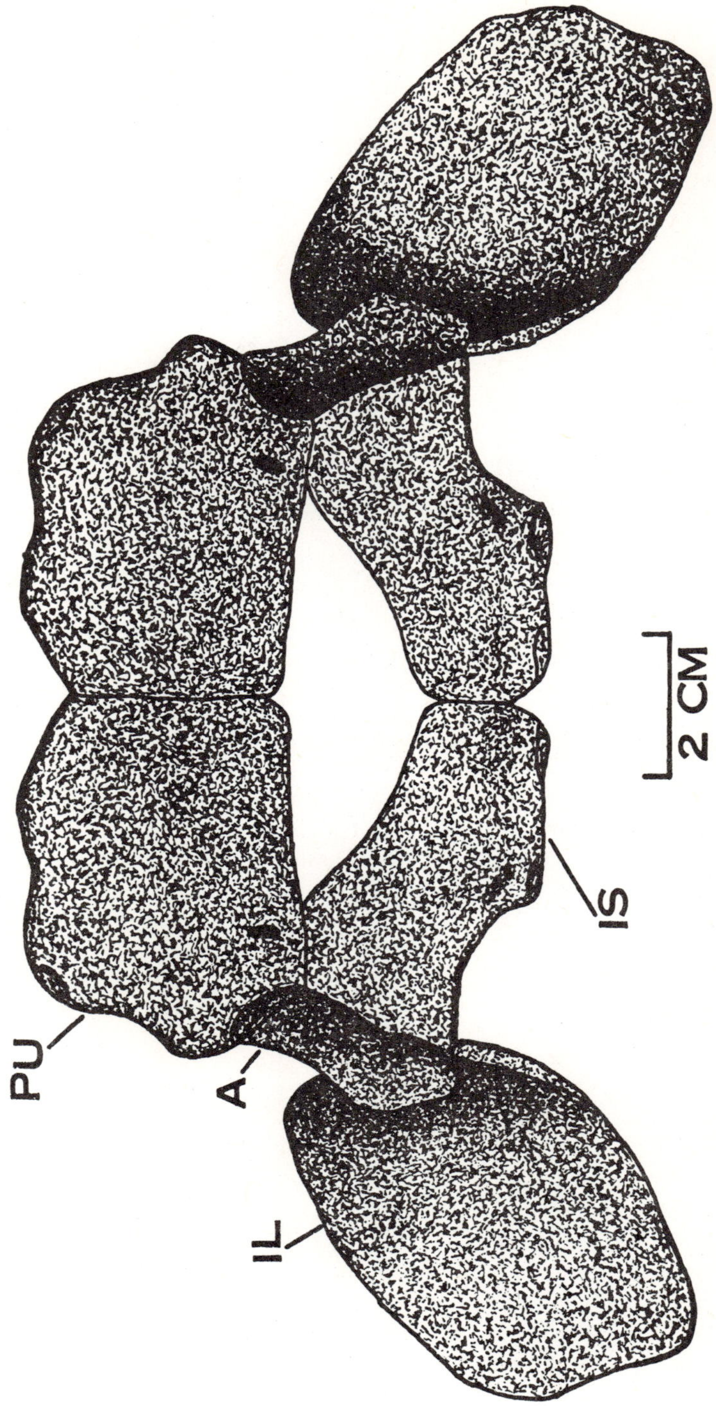

FIG. 20. Ventral view of the pelvis of *Doswellia kaltenbachi*. A—acetabulum, IL—ilium, IS—ischium, PU—pubis.

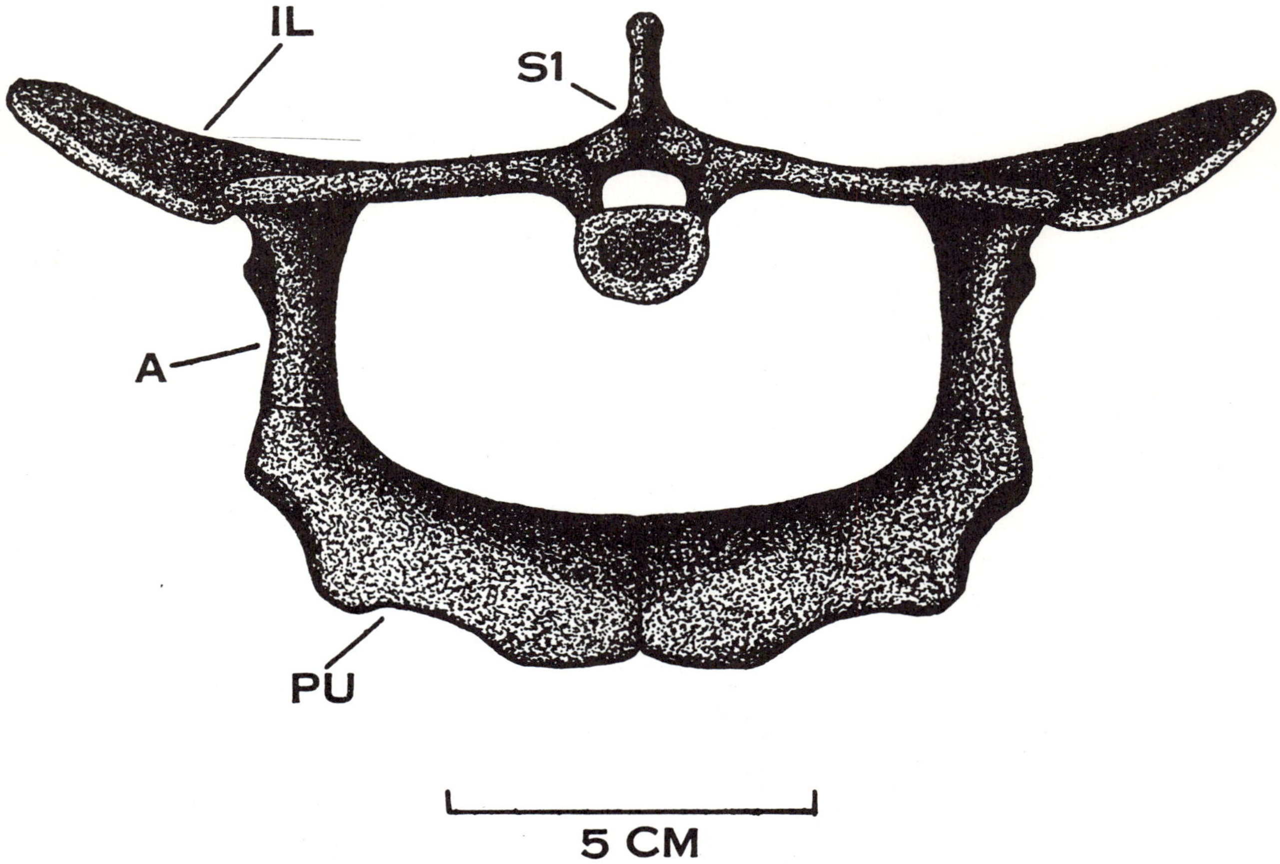

FIG. 21. Anterior view of the pelvis of *Doswellia kaltenbachi.* IL—ilium, PU—pubis, S—sacral vertebra.

tilaginous head. Distally the shaft is slightly swollen. The entire femur appears to have been crushed. It is 11.0 cm long as preserved.

Tibia: Only the distal end of the tibia is known. It has a prominently curved articular surface along which a robust astragulus must have moved.

ARMOR

The armor of *Doswellia* is not readily confused with that of any other well known saurian except possibly that of the aëtosaur *Stagonolepis* (Walker, 1961). A typical plate of *Doswellia* armor is square to rectangular, is sutured to its neighbors laterally (forming transverse rows) and is overlapped by the plate in front of it and overlaps the plate behind it (forming longitudinal columns). Where it is overlapped on its anterior border a distinct and prominent flange is developed which is separated from the rest of the plate by a sulcus. Except for the flange, the surface of each armor plate is covered by scattered pits and bears a low boss which is elongate in the anteroposterior direction and has a papilliform surface.

The pitting and pattern of arrangement is similar to some crocodilian armor, but no crocodile is known to have ever developed *Doswellia*-like flanges marked off by a deep, distinct sulcus; the sulcus is poorly developed or absent in crocodilians. The pitting is similar to that seen in some aëtosaur armor, which likewise is shingled but has only four dorsal and lateral longitudinal columns, never ten columns of plates as is found in *Doswellia.* Parasuchians have shingled armor but the pits are relatively fewer and much larger; the plates vary in different genera from irregularly polygonal to square or rectangular. Unlike *Doswellia,* only two to four longitudinal columns of plates are present. Bipedal thecodonts have never been reported to have more than a single pair of longitudinal columns of armor which do not overlap. Some placodonts were armored but by a mosaic of polygonal pieces sutured together much as in the leather-back turtle, *Dermochelys* (Westphal, 1975); this is unlike the armor of any other Triassic reptile. Thus the armor of *Doswellia* has a unique pattern and arrangement and should be readily distinguishable from that of all other reptiles. Even in the absence of all other skeletal remains individual plates should not be readily confused with anything except aëtosaurs and even here careful attention to the details of pitting pattern may provide a clear distinction.

Although much armor was recovered with the Dos-

FIG. 22. Femur of *Doswellia kaltenbachi.*

well specimen and seven pieces with the Ashland specimen, most of it is so jumbled that it is possible to describe only details of the armor arrangement in the sacral and posterior presacral regions. The pelvic armor includes two longitudinal columns of angled peripheral elements which are sutured anteroposteriorly. These were found near the pelvis and could be reassembled (pl. 8, fig. B); they fit nicely around the curvature of the flared ilia. One of these same elements is also present with the Ashland specimen and its shape is nearly identical to the comparable element in the Doswell specimen. Thus apparently each armor plate in this region had its shape genetically determined and not developmentally determined by growing outward in all directions until it abutted against its neighbors. Ventrally the angled plates end at a smooth edge which indicates that the armor did not extend farther ventro-medially. Dorsally each piece has sutures for attachment to the next more medial longitudinal column. Most of the scattered armor shows sutured edges on the lateral borders but not on the anterior or posterior borders; a few exceptional pieces associated with the pelvis show anterior or posterior suture edges, though no pieces recovered show all edges sutured. Thus a minimum of two transverse rows of medial armor were sutured anteroposteriorly as well as laterally. Since the lateral-most elements of the four pelvic transverse rows are sutured anteroposteriorly, it is likely that the more medial elements of the same transverse rows would likewise be anteroposteriorly sutured. Therefore, although only two sutured transverse rows are directly demonstratable, the presence of four is suspected. Since the pelvic area is largely immobile due to the bracing required for the hips, no motion on the part of the armor is required in this region anyway.

The fourth sutured and angled armor piece in each outer, fused, longitudinal pelvic column probably belongs also to the first caudal transverse row. Each is actually composed of two armor pieces partially fused and partially sutured at the angle. This suggests that all of the other angled, lateral sacral pieces are also actually derived from two separate armor elements. Medial armor plates associated with the pelvis are either about as wide as long or wider than long and the size of the pieces indicates that there were three pairs of longitudinal columns present between the peripheral longitudinal columns. Thus in the sacral region there was probably a base number of five pairs of longitudinal armor columns running down the back and sides of which the outer pairs of longitudinal columns were fused into single, angled elements.

The five-fold base pairs must have continued back some distance onto the tail, but the tail armor is too scattered to be sure how far. One piece of armor with large pits and a strong dorsal boss has a smooth lateral edge and an overall width comparable to that of pieces in the paramedian sacral longitudinal column. The presence of this piece suggests that all longitudinal columns of armor plates on the tail except for the median columns narrowed and pinched out progressively toward the tip of the tail. Once all but the median longitudinal columns were gone, the median pairs then finally began to narrow, becoming a single elongate pair of armor plates along the posterior tail region.

About six transverse rows of armor from the posterior presacral area are partially intact (pl. 9) and offer the only evidence as to the nature of armor details between the skull and hips. Since these were found beneath the articulated string of vertebrae representing dorsals 6–14 they provide strong direct evidence that the armor overlapped toward the posterior end of the animal as should be expected. No transverse row is entirely complete, but two have eight armor plates attached laterally, comprising four matched pairs with the outermost fourth pair having curvature for wrapping around the body from the dorsal to the lateral surface. Since the curved pair terminate laterally at sutured borders, at least a fifth pair of armor plates with smooth and tapered lateral edges must have been present originally. Thus the

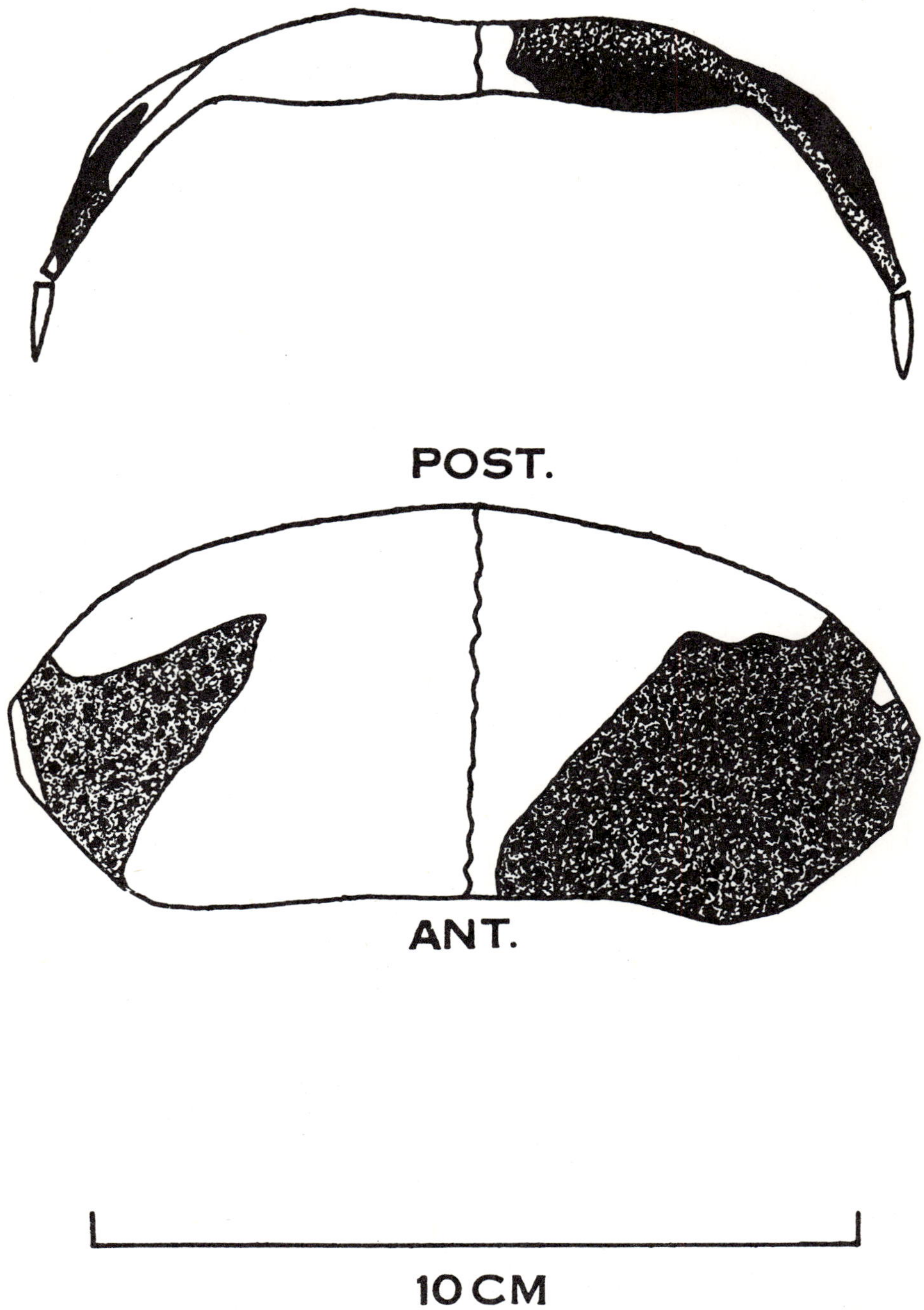

FIG. 23. Anterior and dorsal views of the nuchal collar of *Doswellia kaltenbachi*.

posterior presacral region must have had as many longitudinal columns of armor plates as the sacral region. The median pair of posterior, presacral plates have strong, anteroposteriorly elongate bosses which are located medial to the centerline of each plate. The other pairs have bosses located centrally along the centerline of each plate. Thus the first, fourth, and fifth longitudinal columns of armor have distinctive surface or shape characters, while the second and third are identical with each other though different from the first, fourth and fifth. As in the pelvis, most of the posterior presacral armor is as wide or wider than long, deeply pitted with large pits and keeled with prominent, anteroposteriorly elongate bosses.

In contrast, armor plates associated with the anterior trunk and neck tend to be longer than wide (pl. 8, figs. A, D) with bosses and pits becoming progressively less prominent anteriorly. By the neck region bosses appear to be uniformly low and pitting on the armor plates is only about 1/4 as wide or deep as it is on the posterior trunk armor. The overlapping flanges of the anterior trunk armor plates seem to be relatively longer and thicker than the flanges of either the neck or posterior trunk armor, suggesting that the greatest degree of armor flexibility existed in the shoulder region. A number of strongly curved plates probably come from this region and if so suggest that armor must have extended down the sides of the body. Likewise, several pairs of plates were found in this region which were articulated to each other and obviously came from the same side of the body. Both elements of each pair ended at sutured borders, indicating that, even if one was a paramedian element, minimally a third longitudinal armor column must have been present laterally. In the neck region, an articulated string of three plates from the same row were found. Two appear to be paramedian elements, but the third is not and ends at a sutured border laterally. Thus in the neck as well, an absolute minimum of three pairs of longitudinal armor columns must have been present. Since five pairs of longitudinal armor columns appear to characterize the sacral and posterior dorsal regions, it would not be surprising if the same number of longitudinal columns were present along the anterior trunk and posterior neck as well.

In the Doswell specimen two fragments of armor, uniquely large and thick, were found next to the skull. These are interpreted to be fragments of a pair of enlarged plates that formed a "collar" immediately behind the skull. Their exact placement is equivocal but it is quite likely that they were situated immediately behind the base of the skull and overlying the atlas-axis vertebral complex (fig. 23). The curvature suggests that this plate extended well down the side of the neck, while the flattened ventrolateral edge suggests another small triangular plate carried this protection somewhat farther around. Although of a quite different origin, *Doswellia*'s neck plates form a collar strikingly similar to the nuchal plate of a turtle when viewed anteriorly. Grooves on the dorsal surface of this plate suggest sulci for a mosaic of horny scales on the back of the neck, as though *Doswellia* had a general covering of horny scales over the dermal bony armor.

Numerous scattered pieces having one lateral edge sutured and the other ending at a smooth beveled edge indicate that the armor rows did not extend around the entire girth of the body. Three such elements were found in an articulated column, indicating many of these beveled-edge pieces were aligned one in front of the other. Enough such pieces were found with the Doswell specimen to indicate that a minimum of 23 nonsacral transverse armor rows must have ended in beveled-edged pieces. The incomplete nature of the skeleton indicates that even more may have been present. Moreover, in reptiles generally ventral armor has a different shape or pattern from dorsal armor. Thus there is no reason to suspect that *Doswellia* was also armored ventrally.

TAXONOMIC AFFINITIES

The long and slender stapes unpierced by an arterial foramen, the laterally directed opisthotic, the presence of a suborbital fenestra and the presence of at least an upper temporal opening all clearly place *Doswellia* within the ranks of the sauropsid reptiles. No characters suggest affinity with synapsids, cotylosaurs, or turtles except the primitive ones common to all of these groups, and further comparison can therefore be restricted to archosaurs, lepidosaurs, and "euryapsids." Among these groups, however, the position of *Doswellia* is not so clear. Most of *Doswellia's* striking characters are unique to it and are not useful for allying it to other forms; the remainder are for the most part either primitive for sauropsids, or have appeared independently in quite different lineages.

The salient characters of *Doswellia* are summarized in table 9 under the categories of "unique," "primitive" and "advanced," in the limited context of archosaurian, lepidosaurian and "euryapsid" reptiles. Thus the participation of the surangular in the jaw articular surface is considered to be only an archosaurian trait even though it also occurs in turtles, for the latter group has been excluded from further consideration. For a character to be included in a given group, only some members need possess it to demonstrate that genetic potential exists in that group for acquisition of that particular trait. Thus archosaurs have within different lineages both highly kinetic and akinetic skulls, yet akinesis has been counted as an archosaur trait. Distribution of the characters considered "advanced" in *Doswellia* are indicated by the following letters: A = Archosaurs, L = Lepidosaurs, S = Sau-

TABLE 9

CHARACTER STATES PRESENT IN *Doswellia kaltenbachi*

Primitive Sauropsid Characters by the Late Triassic

1. Pteryygoid teeth retained
2. Quadratojugal braces jaw joint
3. Slender stapes
4. Anterior braincase unossified
5. Suborbital fenestra present
6. Upper temporal opening present
7. Teeth homodont and circular in section
8. No coronoid process on jaw
9. No mandibular foramen
10. No keel on cervicals
11. Amphicoelous vertebral centra
12. Ribs double-headed in neck
13. Caudal ribs retained
14. Clavicles and interclavicle retained
15. T-shaped interclavicle
16. Plate-like ischium and pubis
17. Imperforate acetabulum
18. Femur with internal trochanter and no fourth trochanter

Advanced Sauropsid Characters (Late Triassic and beyond) (numbers are page numbers in Romer, 1956)

1. Strongly akinetic skull (A—142, S—167)
2. Pterygoids around basal braincase (A—139, S—661)
3. Epipterygoids partly cartilaginous (A—p. 130)
4. Quadrate wrapped over columella (T—p. 166)
5. No postfrontal (A—p. 130)
6. Very small or no pineal opening (A—p. 130, T—p. 165)
7. No postparietal (A—p. 130, L—p. 531, S—p. 168)
8. Squamosal "horns" (A—p. 138)
9. Post-temporal fenestrae reduced to small slits or closed (A—p. 138, 156, S—p. 168)
10. Interpterygoid vacuity closed (A—p. 138, S—p. 169)
11. Exoccipital and opisthotic fused (A—p. 137, L—p. 118, S—p. 168, T—p. 166)
12. Skull sculptured (A—p. 139)
13. Skull flattened (A—p. 137, S—p. 168)
14. Fluted teeth (A—*Rutiodon* (*Machaeroprosopus*), S—p. 456)
15. Teeth thecodont (A—p. 443, L—p. 443, S—p. 444)
16. Dentaries form symphysis (A—p. 598, L—p. 207)
17. Long retroarticular process (A—p. 591, L—p. 207, S—p. 668, T—p. 212)
18. No intercentra (A—p. 590, S—p. 659)
19. Increase in sacral vertebrae number (A—p. 288, S—p. 291)
20. Natural break in tail slope (A—p. 272)
21. Ribs change form markedly behind scapula (L—*Icarosaurus*, T—*Trilophosaurus*, S—placodonts)
22. Dorsal ribs distinctly double headed throughout series (A—p. 287)
23. Outflared ilia (A—p. 326, S—*Henodus*)
24. Sigmoidal femur (A—p. 366)
25. Imbricated armor (A—aëtosaurs, some parasuchians)
26. Pitted armor (A—crocodiles, aëtosaurs, parasuchians)

Characters Unique to Doswellia *among Archosaurs*

1. Basipterygoid processes enwrapped by pterygoids
2. Enlarged jugal
3. Prominent supraoccipital crest
4. "Euryapsid" temporal region
5. Square-angled ribs
6. Dorsal rib articulation along ventral groove beneath transverse processes
7. Tail much wider than deep in anterior caudal region

ropterygians and Placodonts, and T = *Trilophosaurus*. When these characters are tallied, 24 are typically present in archosaurs, 14 in sauropterygians and/or placodonts, 5 in lepidosaurs and 5 in *Trilophosaurus*. The preponderance of archosaur characters is emphasized by the peculiarity of a number of these traits to that group: 9 characters are unique to archosaurs, while none are unique to sauropterygians or lepidosaurs and only 1 to *Trilophosaurus*. Thus the characters which appear to be derived clearly indicate that *Doswellia*'s affinities lie with the Archosauria; its ancestor must have been a diapsid and the lack of a lower temporal opening must be secondary.

Following the classification proposed by Romer (1966) the subclass Archosauria includes five orders: Pterosauria, Ornithischia, Saurischia, Crocodilia, and Thecodontia. The Triassic and Early Jurassic ornithischians, saurischians and pterosaurs can be eliminated at once from further comparison with *Doswellia* by virtue of their light build, high and narrow skulls, advanced pelvic structure, and minor or no development of armor. The Triassic and Early Jurassic crocodilians and the Permo-Triassic thecodonts, on the other hand, show many points of similarity with *Doswellia*. Although several suborders of crocodilians are known, they are sufficiently uniform in structure and habitus to permit their comparison as a group to *Doswellia*. The Thecodontia, by way of contrast, include four very diverse suborders (Pseudosuchia, Aëtosauria, Parasuchia, and Proterosuchia) which should be considered separately. One suborder, the Pseudosuchia, is characterized by a high and narrow skull, reduced armor, light body build and advanced pelvic structure paralleling that of the saurischians, ornithischians and pterosaurians; this group likewise contrasts sharply in structure and build with *Doswellia* and need not be considered further. Thus only the Crocodilia, Aëtosauria, Parasuchia and Proterosuchia merit a detailed comparison with *Doswellia*.

Aëtosaurs such as *Typothorax* (Sawin, 1947) and *Stagonolepis* (Walker, 1961) differ from *Doswellia* in that the skull is high, narrow, unsculptured, lacks pterygoid teeth and possesses two large temporal openings. Cervical vertebrae are keeled and there are only two sacrals. The pelvis is strongly triradiate and the ilium is not flared laterally at the top. The tail is high and narrow. While armor is extensively developed and comparably imbricated, it is arranged in only four longitudinal columns dorsally and laterally, never in ten; aëtosaurs are armored ventrally and apparently have no close analogue to *Doswellia*'s "nuchal" plate. Thus *Doswellia*'s similarities to aëtosaurs are superficial and reflect defensive characters acquired by convergence with, rather than intimate relationship to, the aëtosaurs. In the crucial aspects of skull structure and shape and pelvic structure aëtosaurs are notably more advanced than *Doswellia* and their taxonomic affinities lie elsewhere, possibly with the Rauisuchidae (Sill, 1974).

Crocodilians are similar to *Doswellia* in such skull characters as the lack of postfrontals, presence of squamosal "horns," pterygoids which wholly or nearly obliterate the interpterygoid vacuity and are wrapped firmly around the braincase, well developed pterygoid processes on the quadrates and sculptured surfaces. Yet unequivocal crocodilians such as *Protosuchus* (Colbert and Mook, 1951) and *Sphenosuchus* (Walker, 1972) differ in that the lower temporal opening is present, although a few crocodiles of the Late Cretaceous to Recent subfamily Alligatorinae do have the upper opening closed (Romer, 1956, p. 605). Other non-crocodilian characters of *Doswellia* are the lack of a laterosphenoid ossification, vertical quadrate, non-fenestrate lower jaw, location of the retroarticular process far below the articular socket and the role of the surangular as part of the lower jaw articular surface. Especially in the postcranial skeleton *Doswellia* shows strong contrasts with crocodilians. While the dorsally extensive imbricated armor suggests a superficial similarity to crocodilians, in contrast the lack of ventral armor, the lack of ventral keels on the cervicals, the participation of the pubis in the acetabulum, the plate-like pubis and ischium, the laterally outflared iliac blade, the imperforate acetabulum and the presence of three sacral vertebrae (or four if the intimately related first caudal is counted) indicate a profoundly different internal organization from that of crocodilians. The wide and flat tail of *Doswellia,* naturally bent downward at the eighth caudal, also contrasts markedly with the high and narrow sculling tail of crocodilians. Only the marine geosaurs show a natural downward bend to the tail; this is correlated with the development of an ichthyosaur-like swimming tail which is still deep and narrow, not wide and flat. Finally, while crocodilians do have the neck and head of the posterior dorsal ribs nestled beside the transverse processes, they never show the peculiar relationships seen in the posterior dorsals of *Doswellia,* nor is the trunk so sharply divided into anterior and posterior dorsal regions. The crocodilian "lumbar" region is composed only of the last four presacrals; in *Doswellia* the comparable region comprises the last eight presacrals and the appearance is superficially more similar to the trunk of placodonts and *Trilophosaurus* than to crocodilians. Thus the postcranial anatomy of *Doswellia* indicates an arrangement and function quite different from that in any of the recognized suborders of crocodilians.

The large squamosal "horns" at the back of the skull, the elongate snout and the plate-like pelvis might indicate parasuchian affinities for *Doswellia.* Yet since the plate-like pelvis is a primitive character, and squamosal "horns" and long snouts also are present in the clearly non-parasuchian crocodiles and proterochampsids, these three characters are insufficient to provide a strong case for close relationship with the Parasuchia. In other characters no close similarities can be seen. In typical parasuchians such as *Rutiodon* (McGregor, 1906) and *Parasuchus* (Chatterjee, 1978) the postfrontal element is always retained (Romer, 1971), pterygoid teeth are lacking, two temporal openings are always developed, the cervicals are keeled, only two sacral vertebrae are present and the ilia are never outflared. Armor occurs only as one or two pairs of slightly imbricated dorsal longitudinal columns; the tail is high and narrow. All of these characters strongly contrast *Doswellia* with the known genera of parasuchians.

Three major lineages are recognized among the suborder Proterosuchia (Bonaparte, 1971; Sill, 1974), the oldest and most primitive known group of archosaurs. The *Euparkeria-Ornithosuchus* lineage includes bipedal, carnosaur-like animals which have a high and narrow skull, short neck, a fenestrated lower jaw, keeled cervicals and a triradiate pelvis. The latter three character states are clearly more advanced than those found in *Doswellia* and demonstrate that no intimate relationship with *Doswellia* could exist. The second, erythrosuchid-rauisuchid lineage includes only obligatory quadrupedal forms, but otherwise this lineage also contrasts strongly with *Doswellia* in the same traits cited for the *Euparkeria-Ornithosuchus* lineage. Much more similarity exists between *Doswellia* and the third, proterosuchid-proterochampsid lineage. A low-arched skull, unfenestrated jaw, long neck, and plate-like pelvis are all common characters among these forms. All of these traits, however, are primitive for archosaurs (cf. *Proterosuchus* [Cruickshank, 1972]). In the presence of armor, the elaboration of the ilia and sacrals in the pelvis, the specializations of the dorsal and caudal vertebrae and in the various crocodile-like specializations of the skull, *Doswellia* has obviously departed as far or farther from this primitive proterosuchid pattern as the parasuchians, a group also possibly derived from this lineage (Sill, 1974). *Proterosuchus* clearly has the deep and narrow parasuchian-crocodilian type of tail, so *Doswellia* also must have departed farther in its basic lifestyle and locomotor habits from that of the Late Permian-Early Triassic proterosuchids than the parasuchians. Thus placing *Doswellia* in the Proterosuchia would simply make this suborder a "wastebasket" category for structurally primitive archosaurs not readily placed into other categories; the Parasuchia might just as well be retained in the Proterosuchia using such criteria to define this suborder.

The Middle to Late Triassic Family Proterochampsidae (*Proterochampsa, Chanaresuchus, Gualosuchus, Cerritosaurus*) is more advanced than the proterosuchids in skull characters, but no more advanced postcranially. The proper taxonomic position of this group has been disputed. *Proterochampsa* at first was considered to be a primitive crocodile (Sill, 1967), but Walker (1968, 1970) argued that it had more affinity with the Parasuchia than the Crocodilia. More recently Romer (1971) reviewed *Proterochampsa* and the related genera and concluded that they had little to do with either the Crocodilia or Parasuchia; he suggested retaining them in the Proterosuchia as a discrete family, the Proterochampsidae. Sill (1974) has apparently concurred in this latter placement. With their sculptured skulls, elongate snouts, pterygoid teeth and loss of the postfrontals, the proterochampsids show definite similarities to *Doswellia*. Yet the squamosal "horns" are not so nearly well developed, two temporal openings are clearly present in all known proterochampsids, the quadrate does not wrap over the otic notch and no retroarticular process is developed. Furthermore, the jugal borders an antorbital fenestra which, if present in *Doswellia,* certainly was not bordered by this element. No evidence of armor has been reported in these forms (Sill, 1967; Romer,1971). Moreover, the anterior trunk of *Proterochampsa* lacks the elongate cervicals and squared ribs of *Doswellia.* Thus, these forms show none of the postcranial adaptive characters of *Doswellia;* in habitus characters they more nearly parallel the proterosuchians, parasuchians, and crocodilians.

Therefore, although *Doswellia* shows some strong similarities with proterochampsids and crocodilians, and lesser resemblances with parasuchians, these similarities do not provide reason enough to place *Doswellia* in any of these groups as currently defined. No post-Triassic order of archosaurs, except possibly the Crocodilia, and no suborder among the Permo-Triassic order Thecodontia is a suitable category into which *Doswellia* can be placed. Within the order Crocodilia no suborder can be considered to remotely include the sort of postcranial specializations seen in *Doswellia.* This leaves no alternative to the erection of a new suborder to receive this genus.

Whether *Doswellia* should be considered a thecodont or a crocodilian is a tantalizing problem. The very characters which suggest crocodilian affinity for *Doswellia* are the same characters which Sill (1967) originally used to ally *Proterochampsa* with the Crocodilia. Romer (1971) was inclined to believe all of these characters were evolved independently by the proterochampsids and the crocodilians, citing especially the lack of crocodile-like characters in the postcranial skeleton of the proterochampsids (e.g., perforate triradiate pelvis, pubis excluded from the acetabulum, etc.). The same arguments can be cited for *Doswellia.* The basic problem with allying either proterosuchians, proterochampsids, parasuchians or *Doswellia* to crocodilians lies in the necessity of presuming that the strongly triradiate and perforate pelvis of crocodiles was evolved independently of other archosaurs from a primitive plate-like condition without the ancestral forms ever going through a bipedal or terrestrial quadrupedal phase. The high-arched skull of the early crocodilian *Protosuchus* strongly suggests the more likely derivation was from somewhere among the semi-bipedal thecodonts or the progressive quadrupedal erythrosuchid-rauisuchid lineage. Yet to accept this origin implies that the flattened, "horned" and often long-snouted skull of crocodilians evolved parallel to and independently of that found in the proterosuchian-proterochampsid lineage, the parasuchia and in *Doswellia.* Since so many skull characters are similar, but there is no likelihood that these groups are derived from a common ancestor with all of these crocodile-like traits, this requires the assumption that a tremendous amount of parallel evolution has occurred in the skulls of each of these groups.

In the absence of adequate pectoral girdle, wrist, ankle and phalangeal information about *Doswellia,* a definitive decision as to whether *Doswellia* is a thecodont or a crocodile is not yet possible. But for now, since the Thecodontia is essentially a group including animals classified horizontally while the Crocodilia includes (or should include) animals of a specific vertical lineage, it seems best to leave *Doswellia* within the ranks of the Thecodontia. Unless further knowledge of the limbs and girdles of *Doswellia* reveal many more crocodile-like features than are so far evident, *Doswellia's* crocodile-like features are better considered to be a product of parallel evolution. At present, it seems much more likely that *Doswellia* was derived from an animal like *Archosaurus* (Tatarinov, 1961) or *Proterosuchus* (Cruickshank, 1972), and as such ultimately may prove to have somewhat more affinity with the similarly built proterosuchid-proterochampsid lineage as defined by Sill (1974) than with any of the other known groups of thecodonts. The fact that the proterochampsids range into the Late Triassic (Carnian) but show no sign of armor development or other post-cranial specializations approaching those of *Doswellia* debars a simple proterosuchid-proterochampsid-*Doswellia* lineage, however. While the elongate snouts, pterygoid teeth, squamosal "horns" and pelvic structure make an alliance of these forms at least plausible, the wide divergence of *Doswellia* in structure and habitus from these other forms suggests a long independent history for *Doswellia,* perhaps extending back to the Early Triassic or even the Late Permian.

Sill (1974) has tentatively suggested that the Parasuchia have more affinity with the proterosuchids than with the erythrosuchids or ornithosuchids, citing the persistently primitive pelvic structure and the general similarity of body shape. This is certainly a plausible argument, but it should be noted that, unlike proterosuchids, proterochampsids and *Doswellia,* parasuchians have strongly keeled cervicals, a character shared by all more advanced archosaurian lineages. This suggests that the parasuchians probably were derived from an ancestor slightly farther along the main line of archosaurian evolution than the proterosuchids and thus should be more properly allied with the more advanced thecodont lineages. Indeed, the entire proterosuchid-proterochampsid-*Doswellia* group may prove to be so divergent from the rest of the thecodonts that they may collectively prove to merit separate ordinal rank.

The fact that *Doswellia* is both "euryapsid" and an archosaur makes it impossible to argue any longer that the "euryapsid" temporal configuration is a primary indicator of close relationship among all animals possessing this condition. In the case of *Doswellia* it is almost certain that a lower temporal opening was present in its ultimate ancestors even though this opening has been secondarily closed over. If this be true, then other "euryapsids" of less obvious ancestry, especially *Trilophosaurus* (Gregory, 1945) and *Pleurosaurus* (Huene, 1952) could likewise be ultimately of diapsid descent as the rest of their anatomy strongly suggests. Although some "euryapsids" are obviously closely related to each other (e.g., plesiosaurs and nothosaurs) the reasons for establishing this close relationship lie in the presence of numerous other common character states in addition to the presence of a "euryapsid" temporal configuration. Certainly the subclass "Euryapsida" as envisioned by Romer (1956, 1964, 1968) is in large part an artificial concept which is no longer useful except perhaps as a category to contain the nothosaurs, plesiosaurs, and placodonts.

DIAGNOSIS OF THE SUBORDER DOSWELLIINA AND THE FAMILY DOSWELLIIDAE

Although little more than one specimen is currently known, this material is so profoundly distinctive that it clearly merits separate recognition at the familial and subordinal levels. Future discoveries should better indicate which of the generic characters will be also of familial and subordinal value, but for now the following diagnosis is suggested:

Suborder Doswelliina: Skull low and broad, snout elongate. Antorbital vacuity (if present) not bordered by jugal. Orbits high on skull and near midline, "euryapsid" temporal configuration. Postfrontal absent, parietals and frontals laterally narrowed. Occiput wide and low, with no post-temporal fenestrae present. Strongly developed otic notch and squamosal "horns." Interpterygoid vacuity closed and palate flat, no secondary palate developed. Suborbital vacuity present. Pterygoids enwrap basipterygoid. No laterosphenoid ossification. Marginal dentition *ca.* 30–35 teeth in each jaw half, three rows of pterygoid teeth which are rounded and blunt. Anterior dentaries meet along a symphysis, no external mandibular foramen. Prominent retroarticular process. Surangular contributes to articular facet.

Probably twenty-two amphicoelous presacral vertebrae; eight cervicals elongate, cervical ribs short and stout. First six dorsal ribs L-shaped; posterior eight extend laterally but not ventrally, articulate to their respective vertebrae along a groove beneath the transverse processes. Three fused sacral vertebrae, first obviously derived from the presacral series; first caudal vertebra partially incorporated into the sacrum. Anterior caudals bear discrete single-headed ribs. Tail attenuates rapidly within about eight vertebrae along the caudal series to a long "whiplash," eighth caudal along the series has a prominently angled anterior face to produce a natural downward "break" in the tail. Clavicles and interclavicle robust and well ossified. Pubis and ischium flat and plate-like, only slightly downturned at their distal ends. Acetabulum imperforate. Anterior process of iliac blade strongly outflared dorsally to produce an L-shape. Femur sigmoidal with internal trochanter, terminal head, but no fourth trochanter. Stance unimproved to semi-improved quadrupedal. Imbricated armor highly developed dorsally and laterally, unarmored ventrally. Armor over hips sutured rather than imbricated. Extensive gastralia.

Family Doswelliidae: Since there is only one family in the suborder, the familial diagnosis is the same as for the suborder.

FUNCTIONAL ANATOMY

Skull and Jaws: Many characters in the skull of *Doswellia* relate to its strongly akinetic character. The incorporation of the surangular and quadratogugal into the jaw articulation served to help brace and lock the jaw joint, preventing major sliding or slipping along the joint. Only a scissors-like motion was possible, except possibly for a slight fore-aft rocking motion along the smooth proximal end of the quadrate where it meets the squamosal (fig. 11); this would have slightly increased the total mouth gape in a manner similar to that of the early crocodilian *Sphenosuchus* (Walker, 1972). Directly related to the strongly braced jaw joint is the expanded jugal region of the skull and the deep jaw ramus. Expansion of these two regions has afforded space for a large adductor mandibulae complex despite the lack of lower temporal or jaw fenestration. Similarly, the deep retroarticular process and deep posterior jaw margin must have afforded attachment and room for an exceptionally large and powerful depressor mandibulae, proportionately much more powerful than in crocodilians. All of these features suggest jaws which could both clamp shut or pry open with considerable force without danger of dislocating the jaw joint. Corresponding to the strength of the jaws, the quadrates and pterygoids are firmly linked to the skull and the epipterygoids are partially reduced. This indicates (in the absence of an ossified latero-sphenoid) that most stresses in the skull were born by the superficial skull elements. Since the prominent sculpturing found on the surface of comparably akinetic crocodile and stegocephalian skulls is probably a means of reinforcing the superficial bones against stresses generated by the use of their powerful jaw muscles (Coldiron, 1974), this probably best explains why the skull of *Doswellia* is also highly sculptured.

Three characters, the pointed, round and recurved teeth, the elongate jaws and presumably elongate snout, and the largely upward directed eyes, are typically associated with aquatic and piscivorous habits (e.g., nothosaurs, trematosaurs). Teeth of this sort, however, are also common in insectivorous lizards. Moreover, the fact that the pterygoid teeth are round, not sharp, is also anomalous for an obligate fish-eater; sharp teeth to keep fish from escaping while the animal swallowed are the typical development here. Since at least one carnivorous dinosaur had an elongated rostrum (Peter Galton, oral communication, 1977) and at least one skeleton of the long-snouted parasuchian, *Parasuchus,* has been found with a skeleton of the semibipedal thecodont, *Malerisuchus,* in its stomach region (Chatterjee, 1978), long snouts should not always be correlated with piscivorous habits either. Probably the eyes were somewhat "pop-eyed," affording a rather wide field of view and perhaps stereoscopic vision. In normal posture the best vision would have been in an upward direction, but flexion of the neck would have tilted the head downward and thus enhanced forward vision while walking (fig. 25). Thus the placement of the eyes by no means requires the assumption that *Doswellia* was an obligate, aquatic bottom-dweller.

Neck: The elongation of *Doswellia's* neck is pronounced (fig. 24) and the angle of the zygapophyses (*ca.* 40°) suggests considerable flexibility. The prominent squamosal "horns" suggest an exceptionally strong extension was possible by contraction of the epaxial musculature, but some constraint was probably placed on this by the posteriorly imbricated armor, which probably could flex readily in a downward or lateral neck movement but not much above horizontal in an upward movement.

Trunk: The low angle of the dorsal zygapophyses (*ca.* 70° from vertical) suggests that this region of the body flexed readily laterally, but not vertically. The fact that the trunk armor fits neatly over each vertebral segment, and that the vertebrae and ribs are tightly locked by the peculiar system of articulation, further suggests lateral motion was easy while motion in the vertical plane must have been much more difficult. The fact that the armor is imbricated, and that in living armored lizards and crocodilians each armor plate is overlain by a single horny scute, makes it likely that a single horny scute overlay each bony plate and extended posteriorly behind the posterior edge of the bony overlapping flange that covered the anterior border of the next more posterior armor plate. This would have extended protection further across the armor joints and, if sharp, could have afforded a set of posteriorly directed, defensive razors if the back were slightly arched. The shape of the trunk, with its widest portion in the middle, probably served to maximize the visceral volume while at the same time it prevented the body from interfering with the swing of the front and rear limbs during undulatory locomotion.

Tail: The peculiarly bent eighth caudal (fig. 25; pl. 4, fig. M) suggests that the wide, anterior tail region was held parallel to the ground during locomotion, while the remainder of the tail turned sharply down and perhaps dragged the ground. The fact that the prezygapophyses are long and "L"-shaped suggests that the tail had limited lateral flexibility and functioned largely as a semirigid counterweight in walking. The wide, flat and downward-bent shape of the tail would seem to preclude its function as a propulsive organ. Probably when *Doswellia* was alarmed it was capable of pulling the slender portion of its tail between its legs, leaving the flat, broad base as an armor shield to protect the rear of the legs. The development of such a defensive mechanism may have precluded any marked lengthening of the ischium, a character typically seen in most other archosaurs but not in *Doswellia.*

Front Limbs: Although only the clavicles and interclavicle can be reconstructed, their size and build indicates that the front limbs were robust and powerful. No changes in the shape of the interclavicle normally associated with aquatic habits are present. Therefore the front limbs almost certainly functioned as strong walking structures.

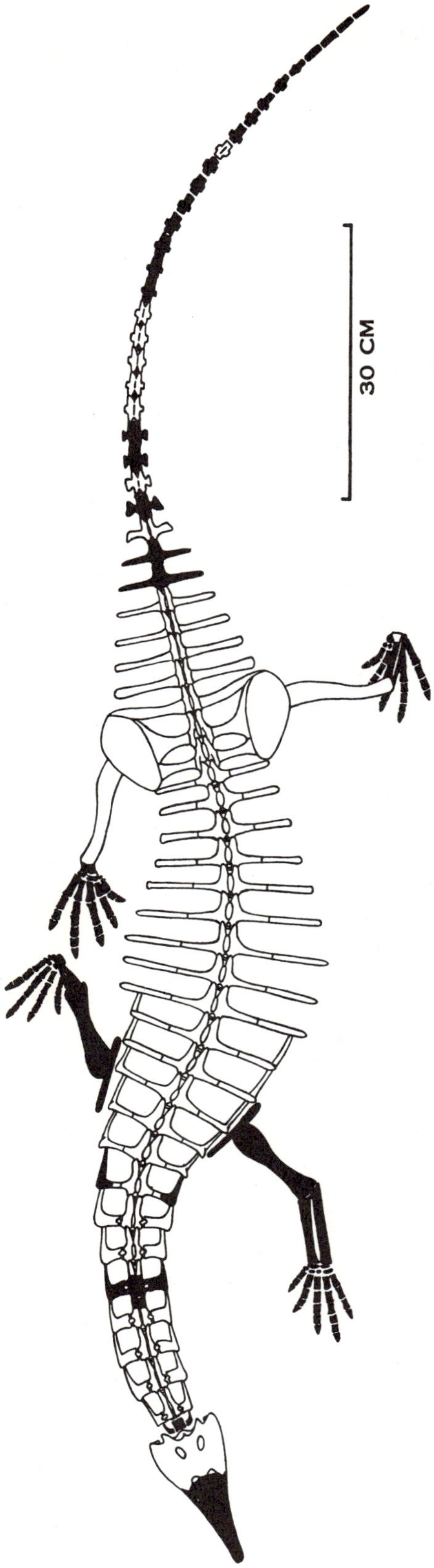

FIG. 24. Dorsal view of the restored skeleton of *Doswellia kaltenbachi.* Dorsal and lateral armor not shown. Black areas inferred.

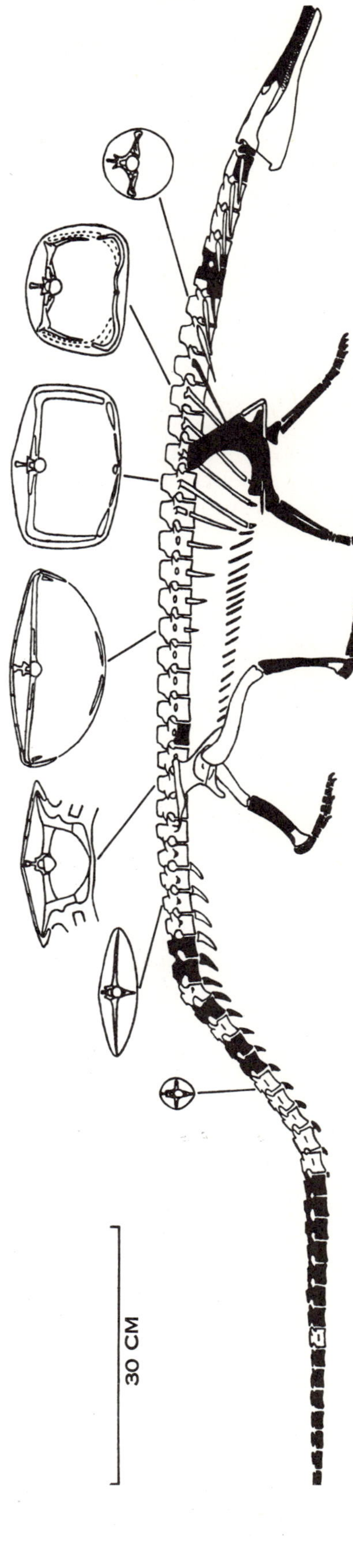

FIG. 25. Lateral view of the restored skeleton of *Doswellia kaltenbachi.* Armor pattern shown in abdominal and pelvic cross-sections where pattern is reasonably well known.

Rear Limbs: The hips are strongly constructed, with three and perhaps four vertebrae modified into a sacrum. These attach to a long iliac blade, which is out-turned much as in ankylosaurs. This is a typical development in heavy graviportal animals and gross analogues exist even in such diverse animals as elephants and giant ground sloths. The strong rim above the acetabulum also argues for a strong hip joint with the femur, so that together these features suggest the presence of powerful rear limbs. The sigmoidal shape of the femur and the fact that the acetabulum is strongly opened downward suggests *Doswellia* might have achieved a near-vertical stance when running, but the plate-like pubis and ischium, non-perforate acetabulum, terminally headed femur and the lack of a fourth trochanter suggest that this apparent capacity was either non-existent or much less well developed than in most of its archosaurian contemporaries. Probably a sprawling to semi-improved posture was much more typical than a running posture (Charig, 1972).

Armor: Although some aspects of the armor have been discussed above, several points remain to be made. The fact that *Doswellia* appears to lack ventral armor suggests that its underbelly was protected partly by gastralia and partly by behavior. Possibly *Doswellia* could roll up in a ball like the cordylid armadillo lizard, but more probably *Doswellia* was capable of digging into and/or clinging to the ground to prevent its underbelly from being exposed. Aardvarks, armadillos, and most lizards of the genus *Cordylus* use similar techniques for protection. Perhaps, like the alligator, *Doswellia* dug holes in mud banks when the weather was unfavorably hot or dry.

Superficially, in its armor coat and elongate snout, *Doswellia* shows certain parallels with aëtosaurs. The internal organization, however, is quite different, and almost certainly we are seeing a case of convergence between two groups of only remotely common ancestry but with a comparable genetic potential. Almost certainly the common ancestor of *Doswellia* and aëtosaurs had none of the common characteristics of armament which both groups displayed by the Upper Triassic.

SPECULATIONS ON THE ECOLOGY OF *DOSWELLIA*

The peculiar combination of many, very primitive, skeletal features with many highly specialized defensive characters such as the strong armor covering, is by analogy very like the sort of character complexes seen in mammals such as the various anteaters, armadillos, and aardvarks and in lizards of the genus *Cordylus*. In the case of the mammals, these animals have become so successful at defense that they have managed to survive successfully at primitive grades of organization relative to their contemporaries and competitors (Vaughn, 1972, pp. 39–42). Generally such animals rely on foraging for food. Although some have specialized food sources, many, especially the species of *Cordylus,* are proficient grubbers and rooters after anything slow enough to be overtaken (Fitzsimons, 1945). Such a lifestyle is envisioned for *Doswellia,* except that it also may have been adept at aquatic living much as are turtles, another armored group. Certainly *Doswellia*'s body build is compatible with a neck lunging piscivorous habit, even while its powerful limb girdles suggest it could move around effectively on land as well. Although its front limbs and rear feet are unknown, *Doswellia's* armored mammalian and reptilian analogues generally have powerful clawed feet and such feet should be anticipated in *Doswellia* both as a means of protecting the soft underbelly and for stirring up food. Clearly the body build does not indicate any proficiency at running down prey or escaping predation by running, though perhaps a brief spurt of speed or a short lunge might have been possible. Rather, foraging or lying in wait were much more likely behaviors.

The actual food source of *Doswellia* is speculative, but several possibilities can be suggested. These are (1) fish, (2) large flying or hopping insects, (3) crayfish, (4) woodboring or secretive crawling insects or insect larvae or (5) bivalves. *Doswellia*'s anatomical specializations would have had similar potential for catching fish or for catching flying or hopping insects. Because of its upward directed eyes *Doswellia* could track anything swimming or flying over its head. With stereoscopic vision to give distance as well as trajectory of the prey, a quick lunge of the neck and head when any animal came close would suffice to catch it. For crayfish, woodboring insects, or secretive crawling insects, the inferred powerful front feet would serve to dig after such prey or break up rotting wood, while the powerful jaws could pry into crayfish holes, under bark, or under rocks to catch the prey. To gather bivalves, *Doswellia* could have used its front feet to unearth bivalves along streambanks or sandbars, the long snout to gather them up and shake away sand and mud between the teeth, and finally the pterygoid teeth to crush the thin shells.

Large flying or hopping insects are totally hypothetical (though hardly unexpected) but all of the other proposed food sources have left some trace of their existence in Newark sediments. Fish scales and casts of thin shelled bivalves are common at the Doswell site where *Doswellia* was collected. A trail of a woodboring creature, presumably an insect larva, was recovered from a petrified log found in the lower Newfound Member only a few hundred stratigraphic feet above the Doswell site, so the presence of creatures of this habitat also is demonstrable. No evidence of crayfish has been so far recognized in the

Doswell Formation, but Paul Olsen (oral communication, 1977) has found abundant evidence of their burrows elsewhere in the Newark Supergroup. Therefore it is highly likely that such creatures existed during Falling Creek time in this area.

In its defensive armament, *Doswellia* strongly resembles certain lizards such as the various species of *Gerrhonotus* and *Cordylus*. These have an armament very similar to that of *Doswellia*. Most *Cordylus* species protect their underbellies largely by clinging to the ground with their strong feet and legs, or lodging themselves under or between rocks and logs while their spike-like and blade-like dermal scutes provide an impregnable dorsal and lateral defense (Fitzsimons, 1945). Almost certainly *Doswellia* employed a similar defensive strategy when caught out in the open and attacked.

Several further inferences about *Doswellia* should be mentioned. These are that the animal was probably secretive, dully colored as a camouflage and "cold blooded." Any animal with a bad taste or poisonous glands would not need *Doswellia*'s armor coat; only a tasty and sought after animal would need this. With all of its other extreme adaptations for defense, it is almost certain that *Doswellia* would also have evolved secretiveness and dull coloration as part of its survival strategy. The probability that *Doswellia* was "cold blooded" is inferred from the fact that structurally it is a primitive archosaur, more primitive than the living crocodilians. Evolutionarily, it lies between a lizard-grade and crocodile-grade of organization and as such probably had a comparable level of physiological organization as well.

Although in many ways quite different, perhaps the best modern analogue of *Doswellia* is the spiny anteater (*Echidna*) of Australia and New Guinea. Here is a long-snouted animal with a spiny protective coat, strong digging ability both to prevent its underbelly from being exposed to attack by carnivores and for food gathering, a browsing habit, primitive locomotive (and reproductive) system, but by parallel evolution an animal which in some ways is quite advanced (e.g., enlarged cerebrum organized quite differently from the placental cerebrum) and specialized (no teeth, long tongue, spiny coat, etc.). *Doswellia*, although different in many other features, was similarly a "living fossil" in its own time and was a strange animal even by the standards of the Triassic. Certainly in many ways it was an armored relict of an earlier, perhaps Permian, radiation of the archosaurian reptiles which by its peculiar specializations managed to survive in the highly competitive and hostile world of the upper Triassic parasuchians and dinosaurs.

ACKNOWLEDGMENTS

Assistance was rendered in excavating the materials described in this paper from the Doswell site by the following individuals: Mr. Rudolph Bland, Jr., Mr. James Kaltenbach, Ms. Marguerite Kingston, Ms. Siobhan LaPiana, Ms. Suzanne LaPiana, Mr. Vincent LaPiana, Mr. Gus McCracken, Col. Thomas McCracken (Ret.), Mr. Peter McCrery, Ms. Margaret Miller, Dr. W. Schyler Miller, Mr. W. Schyler Miller, Jr., Mr. Bobby Neal, Ms. Val Neal, Mr. W. T. (Bill) Neal and Mr. Don Woolford. Dr. J. Buel Weems took photographs of the excavation site which were critical in reconstructing the detailed section of the Doswell site. Mr. Troy Ledbetter of the Ledbetter Construction Company kindly gave permission for the excavation to be conducted and Mr. Luke Ledbetter of that company kindly dislodged a number of large blocks with a backhoe, for later hand reduction. Mr. Donald Beazely donated a block of matrix which proved to contain the posterior portion of the skull of *Doswellia*, and the ultimate description of this genus would have suffered considerably without his generosity. Mr. Arnold Lewis of the U.S. National Museum offered invaluable advice concerning laboratory preparation of the specimens.

Principal advice during preparation of the manuscript was provided by Dr. Anthony G. Coates, Dr. Nicholas Hotton III, and Dr. Roy Lindholm, while valuable discussion and assistance were also rendered by Dr. Donald Baird, Dr. Sankar Chatterjee and Dr. George Zug.

Assistance in preparation of the manuscript was graciously provided by Ms. Elise Giarimoustas, Ms. Marguerite LaPiana, Ms. Brenda Livsey-Coates and Ms. Jess Muelken. Mr. Vincent LaPiana helped with the preparation of the figures and Dr. Anthony G. Coates assisted in the preparation of the plates. Dr. Peter Dodson made final review of the manuscript for the American Philosophical Society.

Without the help and constructive criticisms of all of these people, this paper would have been completed much more slowly or not at all. The author wishes to express a deep sense of gratitude to each of them for their contribution to the successful completion of this study.

REFERENCES

Blatt, H., G. Middleton, and R. Murray, 1972. *Origin of Sedimentary Rocks* (Prentice-Hall Inc., Englewood Cliffs, New Jersey), 634 pp.

Bonaparte, J., 1971. *Los Tetrápods del Sector Superior Formación Los Colorados, la Rioja, Argentina (Triássico Superior)*, pt. 1: Opera Lillona **22**: pp. 1–183.

Charig, A. J., 1972. "The evolution of the archosaur pelvis and hind limb; an explanation in functional terms," in *Studies in Vertebrate Evolution* (Winchester Press, New York), pp. 121–155.

Chatterjee, S., 1978. "A primitive parasuchid (phytosaur) reptile from the Upper Triassic Maleri Formation of India," *Palaeontol.* **21**(1): pp. 83–217.

Colbert, E. H., and C. C. Mook, 1951. "The ancestral crocodilian *Protosuchus*, *Bull. American Mus. Nat. Hist.* **97**(3): pp. 143–182.

Coldiron, R. W., 1974. "Possible functions of ornament in labyrinthodont amphibians," Occasional Papers of the Museum of Natural History of the University of Kansas **33**: pp. 1–9.

Cornet, B., 1977. "The palynostratigraphy and age of the Newark Supergroup," Ph.D. thesis, the Pennsylvania State Univ., University Park, Pennsyl., 505 pp.

Cruickshank, A. R. I., 1972. "The proterosuchian thecodonts," in *Studies in Vertebrate Evolution,* K. A. Joysey and T. S. Kemp, ed. (Oliver and Boyd, Edinburgh), pp. 89–119.

Dunay, R. E., and M. J. Fisher, 1974. "Late Triassic palynofloras of North America and their European correlatives," in *Permian and Triassic Palynology: Rev. Palaeobot. Palynol.* **17**(1–2) : pp. 179–186.

Emmons, E. E., 1856. *Geological Report on the Midland Counties of North Carolina.* (G. P. Putnam & Co., Raleigh, N.C.), 352 pp.

Emmons, E. E., 1857. *American Geology, containing a statement of the principles of the science, with full illustrations of the characteristic American fossils, with an atlas and a geological map of the United States.* (Albany, N.Y.), part 6, 152 pp.

Fitzsimons, V. F., 1945. *Lizards of South Africa.* (Voortrekkerpers, Beperk, Johannesburg, U. S. A.), 437 pp.

Fontaine, W. M., 1883, "Contributions to the knowledge of the older Mesozoic flora of Virginia," *United States Geol. Survey Mon.* **6**: pp. 1–144.

Gregory, J. T., 1945. "Osteology and relationships of *Trilophosaurus,*" *Univ. of Texas Publ.* **4401**: pp. 273–359.

Huene, F., 1952. "Revision der Gattung *Pleurosaurus* auf Grund neuer und alter Funde," *Palaeontographica,* A **101**: pp. 167–200.

Knowlton, F. H., 1899. "Report on some fossil wood from the Richmond Basin, Virginia," *United States Geol. Survey Annual Rept.* **19**(2) : pp. 516–519.

McGregor, J. H., 1906. "The Phytosauria, with especial reference to *Mystriosuchus* and *Rhytidodon,*" *Mem. American Mus. Nat. Hist.* **9**: pp. 29–101.

Olsen, P., 1978. "On the use of the term Newark for Triassic and Early Jurassic rocks of eastern North America," Newsl. Stratigr. 7(2) : pp. 90–95.

Riley, J. P., and R. Chester, 1971. "Introduction to Marine Geochemistry" (Academic Press, New York), 465 pp.

Romer, A. S., 1956. *Osteology of the Reptiles.* (University of Chicago Press, Chicago), 772 pp.

——. 1966. *Vertebrate Paleontology* (3rd ed.). (University of Chicago Press, Chicago), 468 pp.

——. 1968. *Notes and Comments on Vertebrate Paleontology.* (University of Chicago Press, Chicago), 304 pp.

——. 1971. "The Chañares (Argentina) Triassic Reptile Fauna; XI, Two new long-snouted thecodonts, *Chanaresuchus* and *Gualosuchus,*" *Breviora* **379**: pp. 1–22.

Sawin, H. J., 1947. "The pseudosuchian reptile *Typothorax meadei,* new species," *Jour. Paleontology* **21**(3) : pp. 201–238.

Schaeffer, B., and N. G. McDonald, 1978. "Redfieldiid fishes from the Triassic-Liassic Newark Supergroup of Eastern North America," *Bull. American Mus. Nat. Hist.* **159**(4) : pp. 129–174.

Sill, W. D., 1967. "*Proterochampsa barrionuevoi* and the early evolution of the Crocodilia," *Bull. Mus. Comparative Zool.* **135**: pp. 415–446.

——. 1974. "The anatomy of *Saurosuchus galilei* and the relationships of the rauisuchid thecodonts (Carnian of Argentina)," *Bull. Mus. Comparative Zool.* **146**(7) : pp. 317–362.

Tatarinov, L. P., 1961. "Material on the U.S.S.R. pseudosuchians, *Palaeont. Zh. U.S.S.R.* **1**: pp. 117–132.

Vaughn, T. A., 1972. *Mammalogy.* (W. B. Saunders Co., Philadelphia), 463 pp.

Walker, A. D., 1961. "Triassic reptiles from the Elgin area: *Stagonolepis, Dasygnathus* and their allies." *Phil. Trans. Royal Soc. London* **B244**: pp. 103–204.

——. 1968. "*Protosuchus, Proterochampsa* and the origin of phytosaurs and crocodiles," *Geol. Mag.* **105**: pp. 1–14.

——. 1970. "A revision of the Jurassic reptile *Hallopus victor* (Marsh) with remarks on the classification of crocodiles," *Roy Soc. London, Phil. Trans.,* Ser. B, **257**(816) : pp. 323–371.

——. 1972. "New light on the origin of birds and crocodiles," *Nature* **237**: pp. 257–263.

Weems, R. E., 1980. "Geology of the Taylorsville Basin," Virginia Div. Min. Resources, Contributions to Virginia Geol. **4**: (in press).

Westphal, F., 1975. "Principles of structure and growth in the dermal armor of placodonts (Reptilia triadica)," *Paläntol. Zeit.* **49**(1–2) : pp. 97–125.

PLATE 1. A—Dorsal view of isolated jugal element of *Doswellia kaltenbachi* from the Doswell site. B—Ventral view of same element in A. C—Dorsal view of skull of *Doswellia kaltenbachi* (USNM 214823).

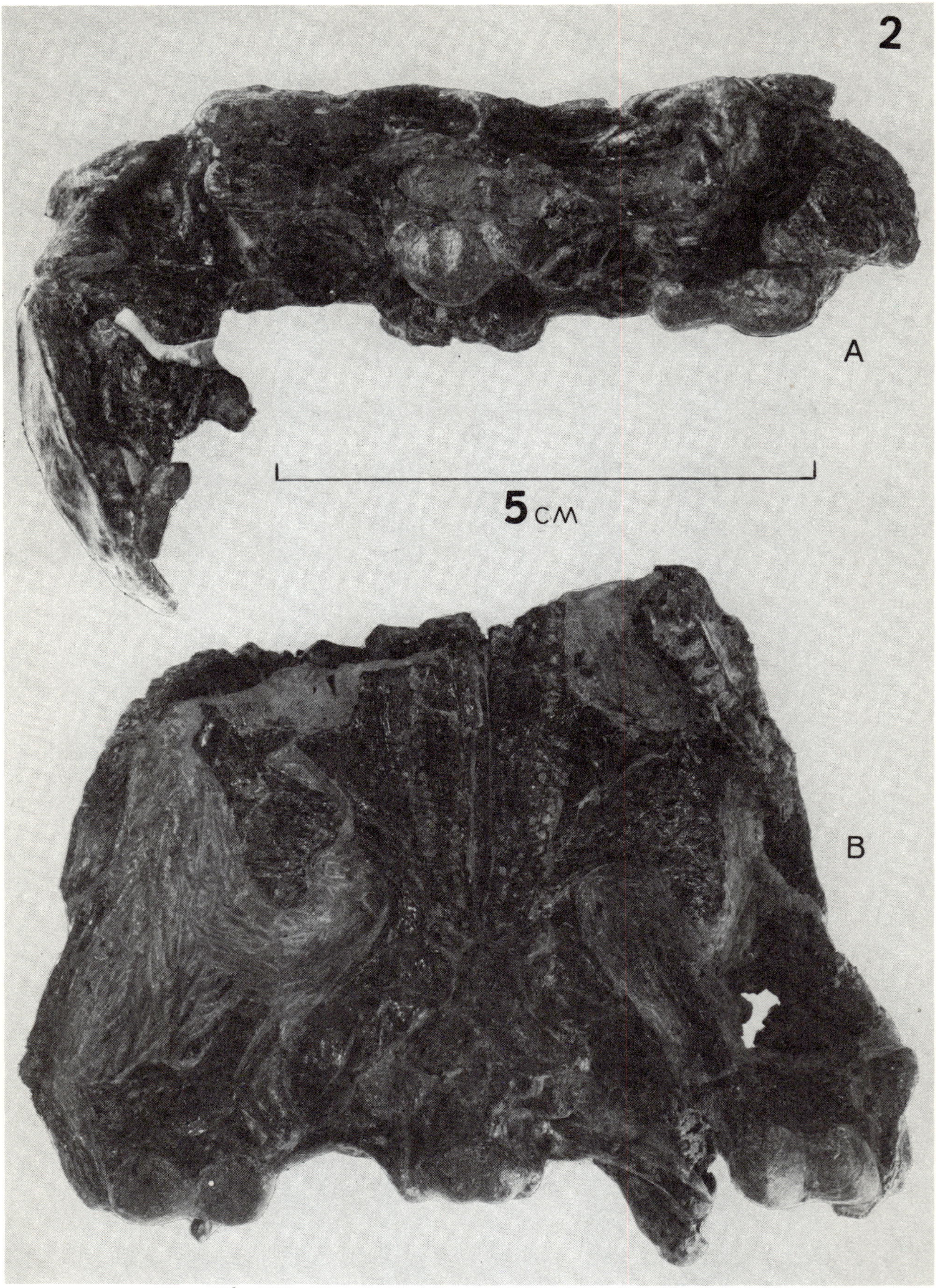

PLATE 2. A—Occipital view of the skull and left jaw of *Doswellia kaltenbachi* (USNM 214823). B—Ventral view of the skull of *Doswellia kaltenbachi* (USNM 214823).

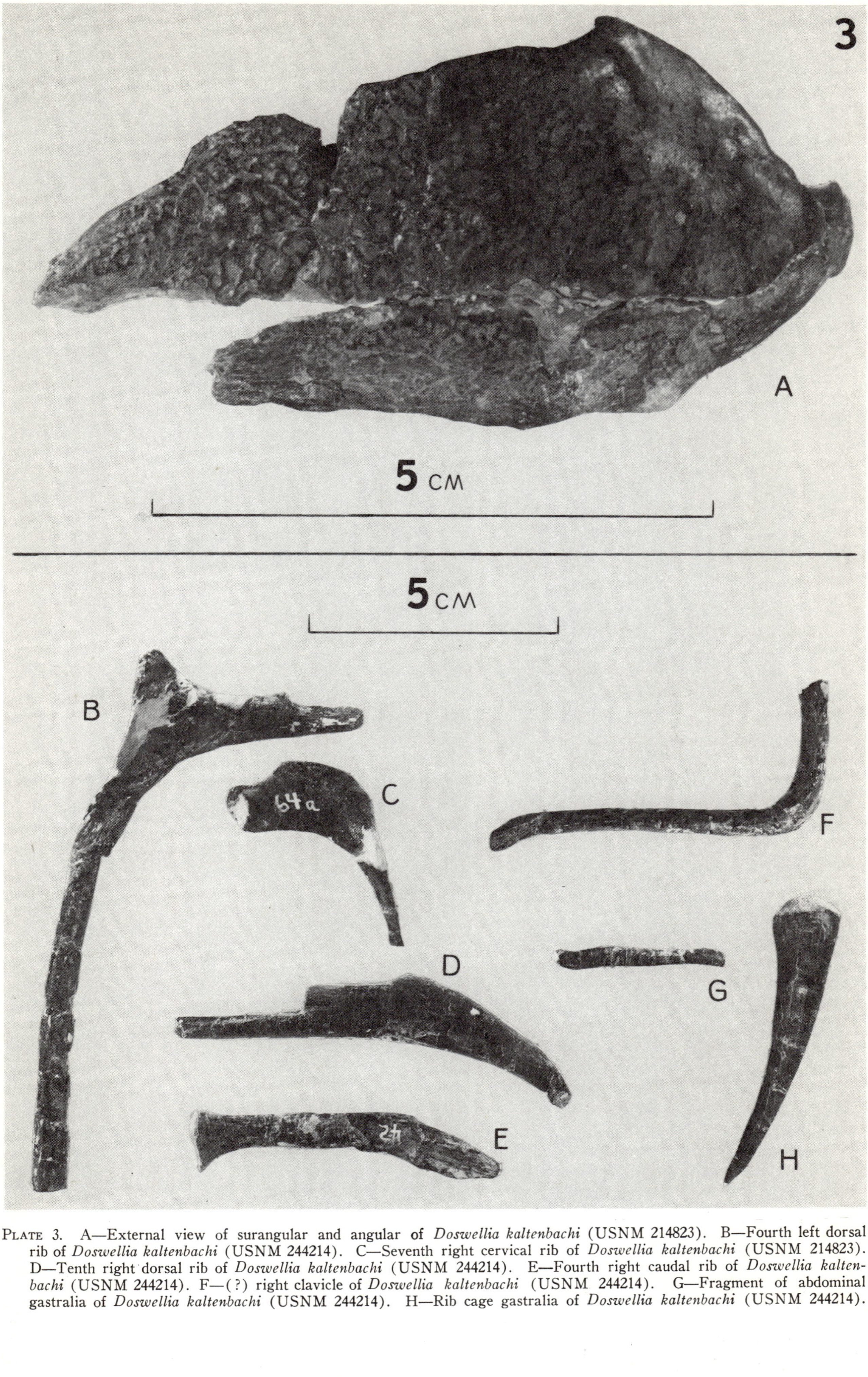

PLATE 3. A—External view of surangular and angular of *Doswellia kaltenbachi* (USNM 214823). B—Fourth left dorsal rib of *Doswellia kaltenbachi* (USNM 244214). C—Seventh right cervical rib of *Doswellia kaltenbachi* (USNM 214823). D—Tenth right dorsal rib of *Doswellia kaltenbachi* (USNM 244214). E—Fourth right caudal rib of *Doswellia kaltenbachi* (USNM 244214). F—(?) right clavicle of *Doswellia kaltenbachi* (USNM 244214). G—Fragment of abdominal gastralia of *Doswellia kaltenbachi* (USNM 244214). H—Rib cage gastralia of *Doswellia kaltenbachi* (USNM 244214).

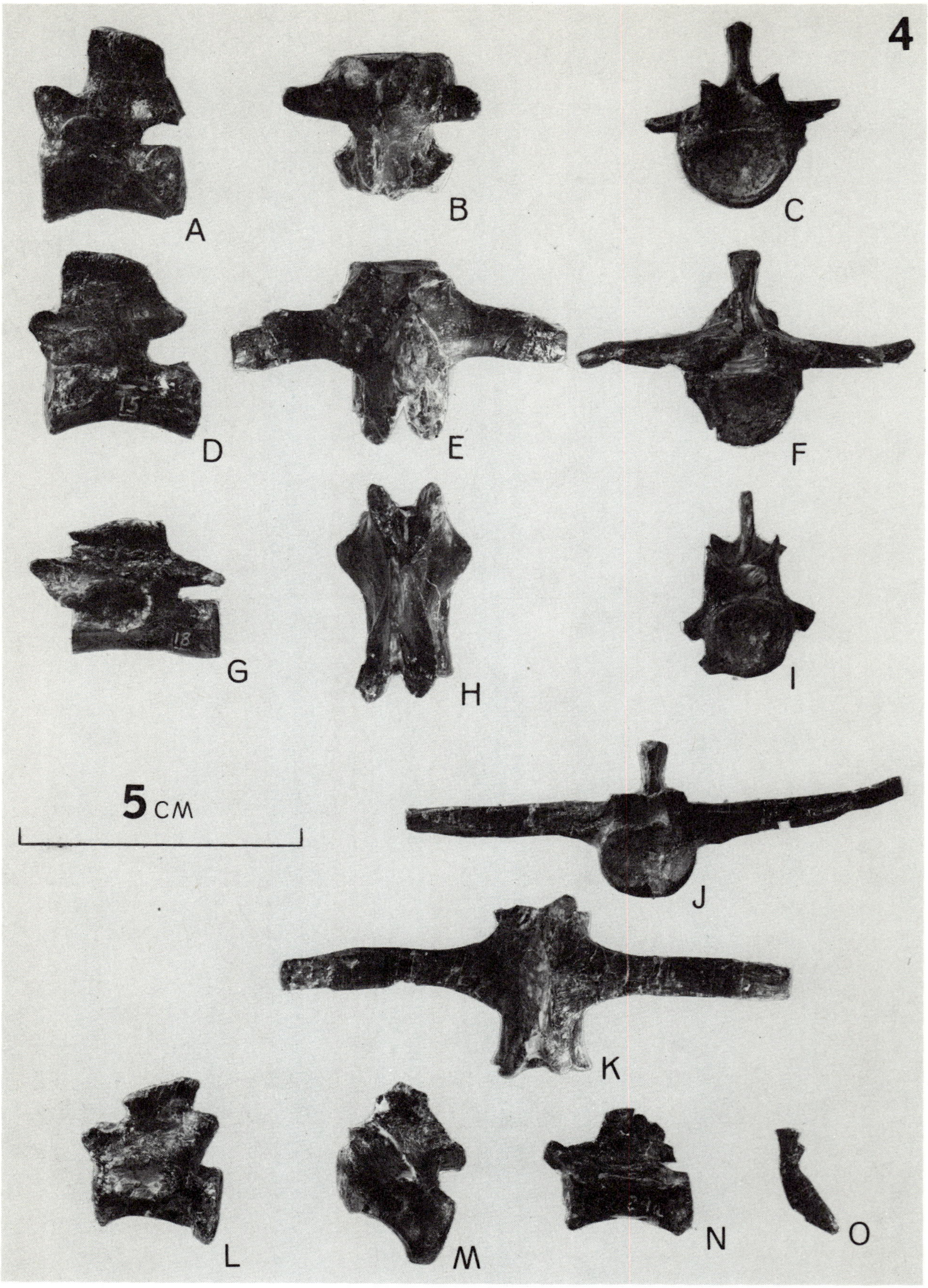

PLATE 4. A—Lateral view of first sacral vertebra of *Doswellia kaltenbachi* (USNM 244214). B—Dorsal view of first sacral vertebra of *Doswellia kaltenbachi* (USNM 244214). C—Anterior view of first sacral vertebra of *Doswellia kaltenbachi* (USNM 244214). D—Lateral view of second sacral vertebra of *Doswellia kaltenbachi* (USNM 244214). E—Dorsal view of fourth dorsal vertebra of *Doswellia kaltenbachi* (USNM 244214). F—Anterior view of fourth dorsal vertebra of *Doswellia kaltenbachi* (USNM 244214). G—Lateral view of third sacral vertebra of *Doswellia kaltenbachi* (USNM 244214). H—Dorsal view of seventh cervical vertebra of *Doswellia kaltenbachi* (USNM 244214). I—Anterior view of seventh cervical vertebra of *Doswellia kaltenbachi* (USNM 244214). J—Anterior view of tenth dorsal vertebra of *Doswellia kaltenbachi* (USNM 244214). K—Dorsal view of tenth dorsal vertebra of *Doswellia kaltenbachi* (USNM 244214). L—Lateral view of fourth caudal vertebra of *Doswellia kaltenbachi* (USNM 244214). M—Lateral view of eighth caudal vertebra of *Doswellia kaltenbachi* (USNM 244214) showing angled anterior central face. N—Lateral view of (?)fifteenth caudal vertebra of *Doswellia kaltenbachi* (USNM 244214). O—Lateral view of a chevron of *Doswellia kaltenbachi* (USNM 244214).

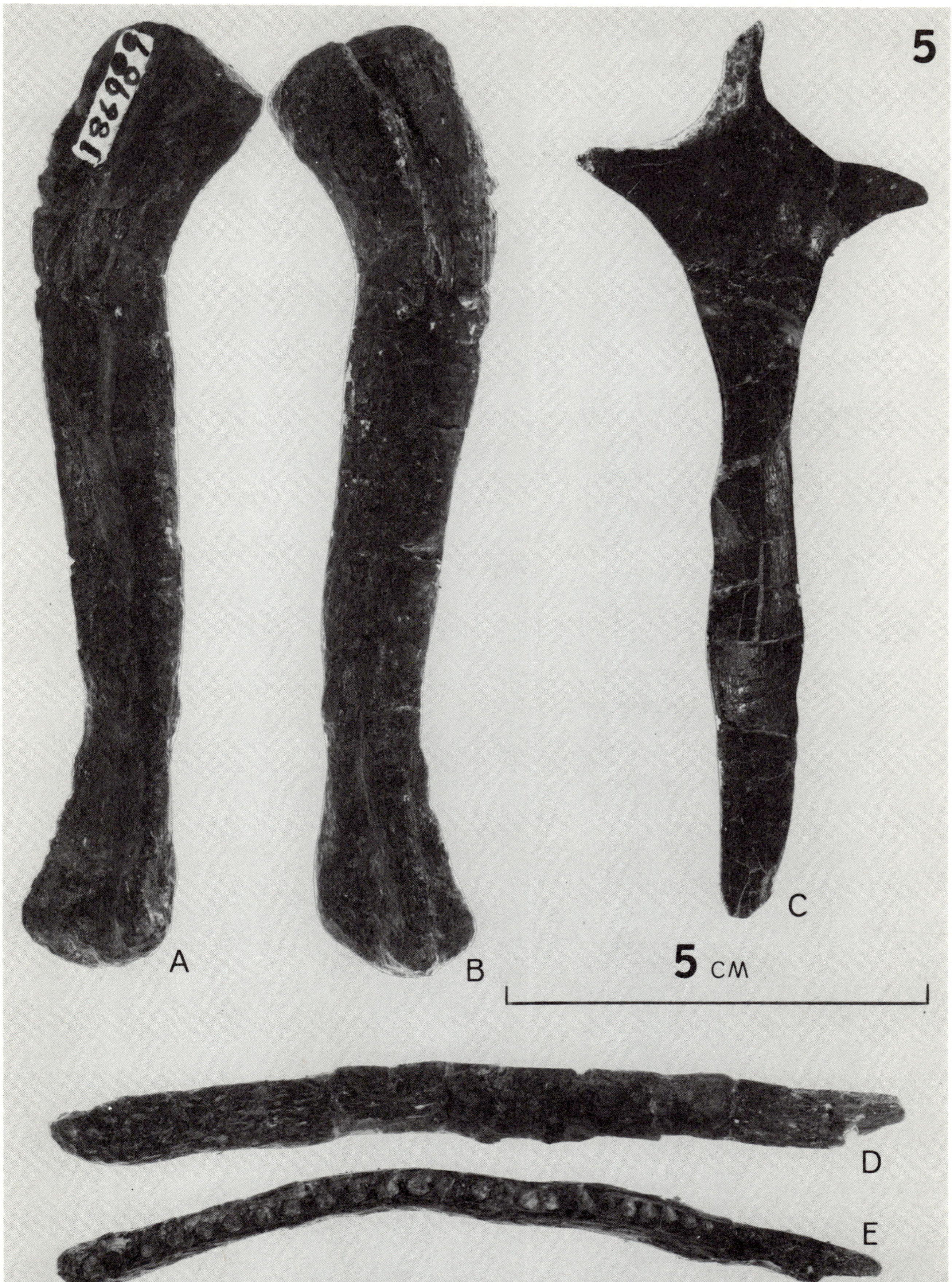

PLATE 5. A—External view of left femur of *Doswellia kaltenbachi* (USNM 186989). B—Internal view of femur shown in A. C—External view of interclavicle of *Doswellia kaltenbachi* (USNM 244214). D—External view of dentary of *Doswellia kaltenbachi* (USNM 186989). E—Dorsal view of dentary shown in D.

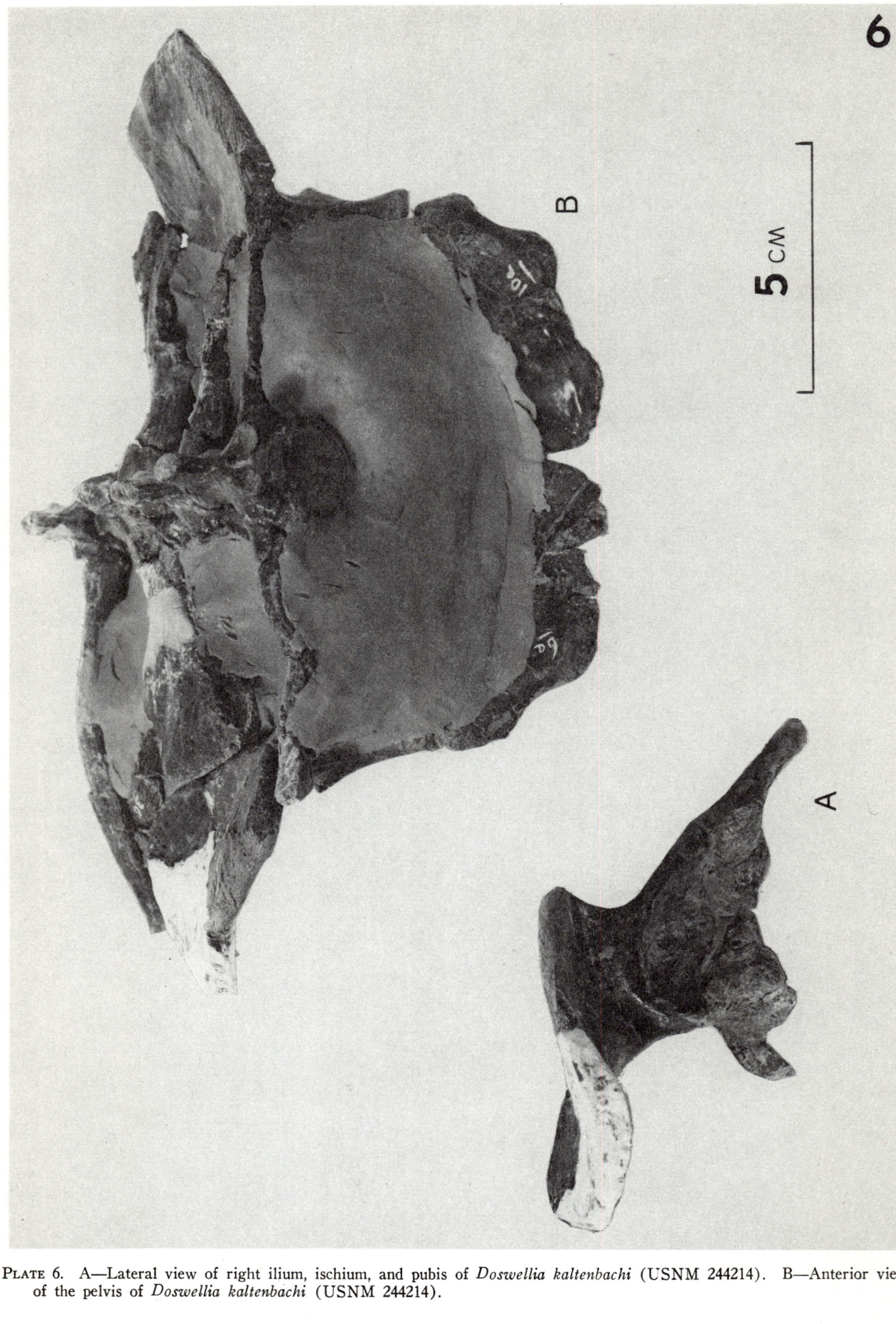

PLATE 6. A—Lateral view of right ilium, ischium, and pubis of *Doswellia kaltenbachi* (USNM 244214). B—Anterior view of the pelvis of *Doswellia kaltenbachi* (USNM 244214).

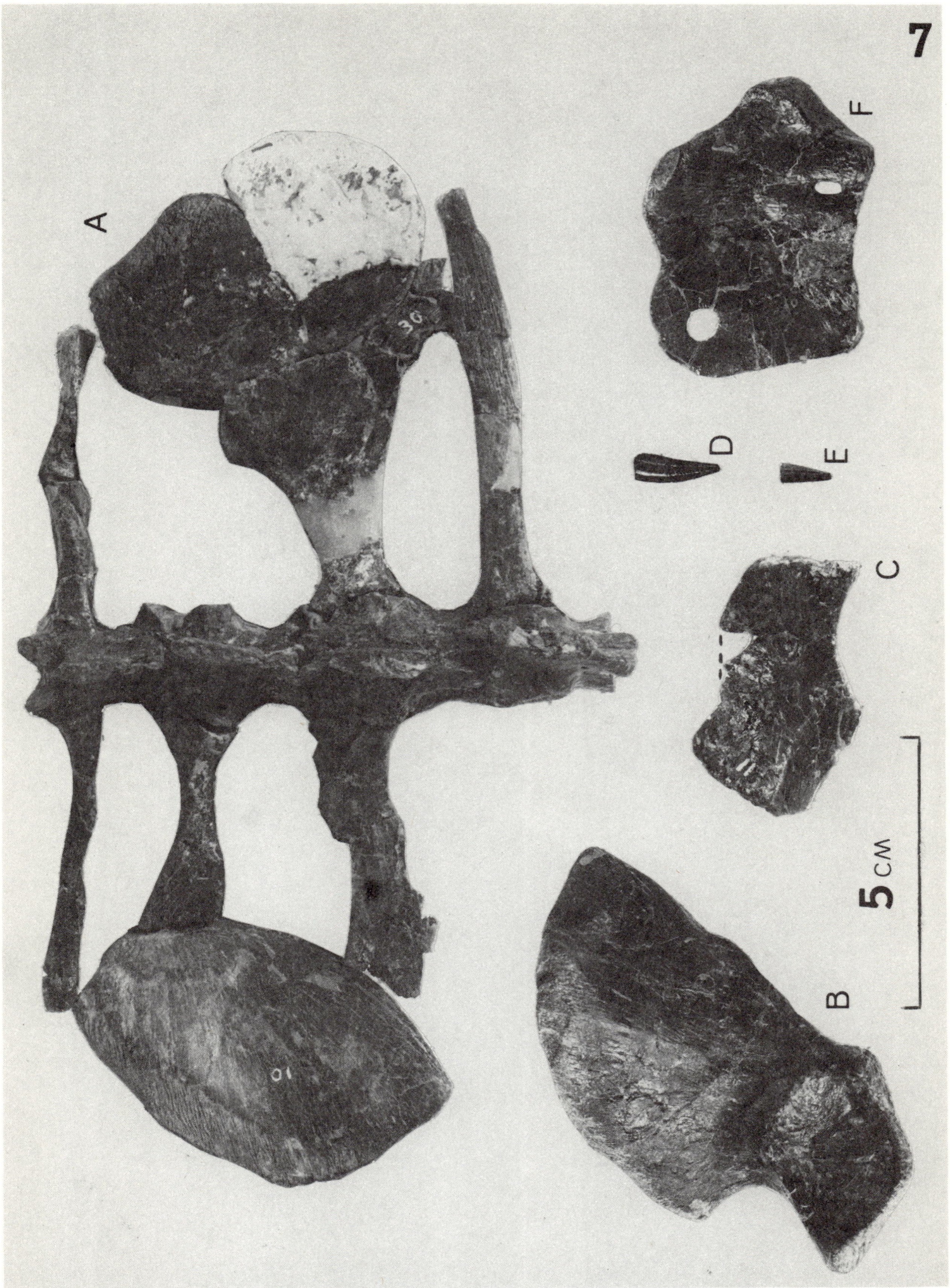

PLATE 7. A—Dorsal view of pelvis of *Doswellia kaltenbachi* (USNM 244214). B—External view of right ilium of *Doswellia kaltenbachi* (USNM 244214). C—External view of right ischium of *Doswellia kaltenbachi* (USNM 244214). D—Lateral view of anterior parasuchian tooth (*cf. Rutiodon carolinensis*) from bed 8 of the Doswell site (see figs. 4 and 5). E—Lateral view of tooth tentatively assigned to *Doswellia kaltenbachi* from bed 8 of the Doswell site. F—External view of right pubis of *Doswellia kaltenbachi*. Note puncture (upper left) probably made by a parasuchian tooth such as the one illustrated in D.

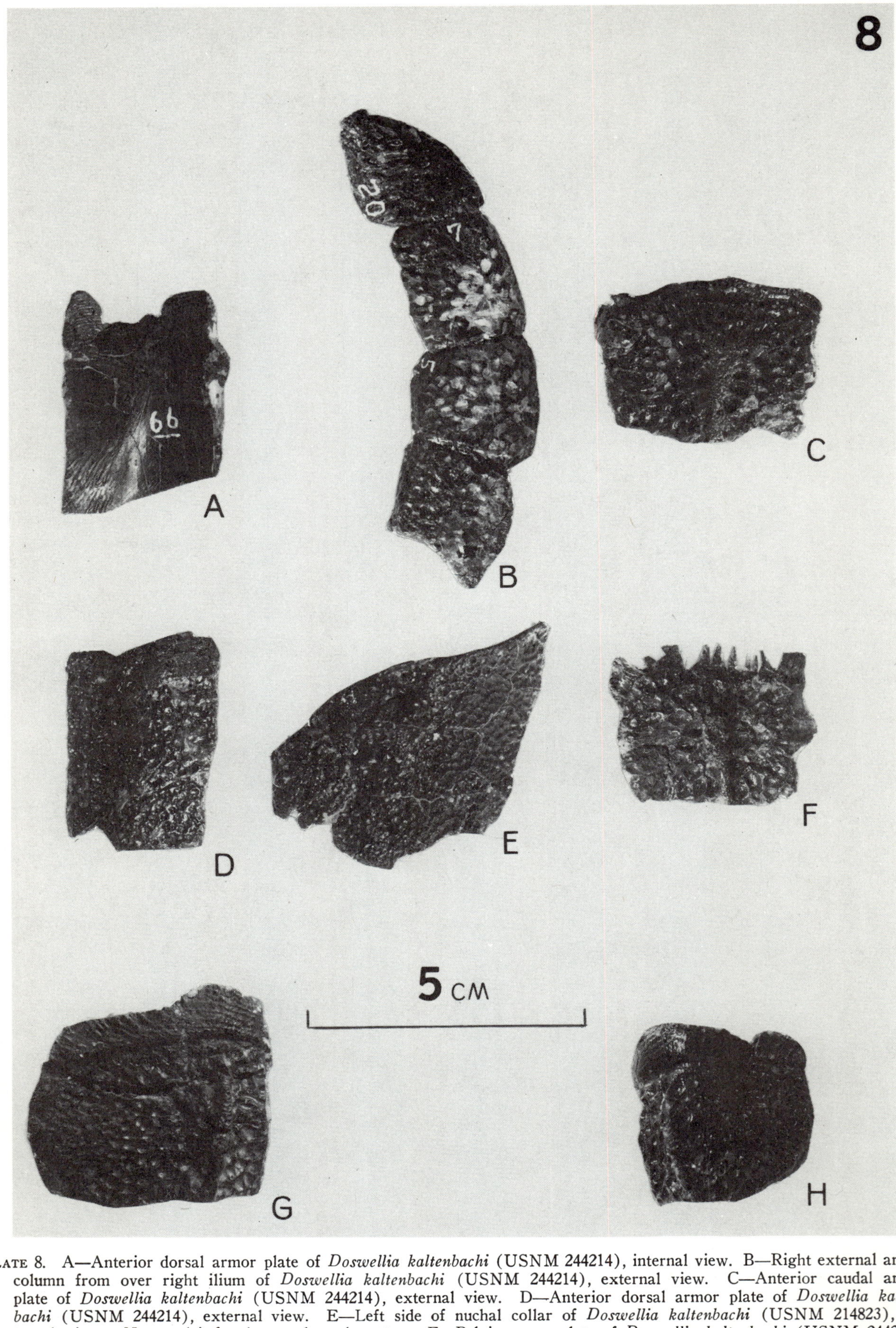

PLATE 8. A—Anterior dorsal armor plate of *Doswellia kaltenbachi* (USNM 244214), internal view. B—Right external armor column from over right ilium of *Doswellia kaltenbachi* (USNM 244214), external view. C—Anterior caudal armor plate of *Doswellia kaltenbachi* (USNM 244214), external view. D—Anterior dorsal armor plate of *Doswellia kaltenbachi* (USNM 244214), external view. E—Left side of nuchal collar of *Doswellia kaltenbachi* (USNM 214823), external view. Note sulci for horny dermal scutes. F—Pelvic armor plate of *Doswellia kaltenbachi* (USNM 244214), external view. G—Left external (?)mid-dorsal armor plate of *Doswellia kaltenbachi* (USNM 244214). Left side ends in beveled edge, right side has sutures. H—Angled (?)posterior dorsal armor plate of *Doswellia kaltenbachi* (USNM 244214).

PLATE 9. Semi-articulated patch of *Doswellia* armor from the posterior presacral region (USNM 244214). Most complete row (between arrows) has four matched pairs of plates. Both sides end in sutured borders, indicating at least one more pair of plates were originally present. Plates overlap posteriorly (up is anterior).

10

PLATE 10. Block containing type specimen of *Doswellia kaltenbachi* (USNM 244214) before disassembly. See Figure 7 for comparison.

PLATE 11. Closer view of articulated vertebral string in Plate 10. Note grooves on ventral side of transverse processes for reception of rib neck and heads.

Oversize 060 Am35t v.70 pt.7

Weems, Robert E.

An unusual newly discovered archosaur from the Upper